CAMBRIDGE STUDIES
IN MATHEMATICAL BIOLOGY: 7
Editors
C. CANNINGS
Department of Probability and Statistics, University of Sheffield,
Sheffield, U.K.
F. C. HOPPENSTEADT
College of Natural Sciences, Michigan State University,
East Lansing, Michigan, U.S.A.
L. A. SEGEL
Weizmann Institute of Science, Rehovot, Israel

MATHEMATICAL ASPECTS
OF HODGKIN-HUXLEY
NEURAL THEORY

JANE CRONIN

Professor of Mathematics, Rutgers University

Mathematical aspects of Hodgkin–Huxley neural theory

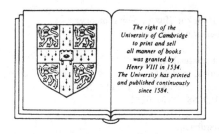

The right of the
University of Cambridge
to print and sell
all manner of books
was granted by
Henry VIII in 1534.
The University has printed
and published continuously
since 1584.

CAMBRIDGE UNIVERSITY PRESS

Cambridge

New York   New Rochelle   Melbourne   Sydney

CAMBRIDGE UNIVERSITY PRESS
Cambridge, New York, Melbourne, Madrid, Cape Town, Singapore, São Paulo

Cambridge University Press
The Edinburgh Building, Cambridge CB2 8RU, UK

Published in the United States of America by Cambridge University Press, New York

www.cambridge.org
Information on this title: www.cambridge.org/9780521334822

First published 1987
This digitally printed version 2008

A catalogue record for this publication is available from the British Library

Library of Congress Cataloguing in Publication data

Cronin, Jane, 1922–
  Mathematical aspects of Hodgkin–Huxley neural theory.
(Cambridge studies in mathematical biology; 7)
Bibliography: p.
1. Neural conduction—Mathematical models.
2. Purkinje cells—Mathematical models.
3. Electrophysiology—Mathematical Models. I. Title. II. Series
QP363.C76 1987   599'.0188   87-9241

ISBN 978-0-521-33482-2 hardback
ISBN 978-0-521-06388-3 paperback

CONTENTS

CONTENTS

PREFACE

The quantitative study of electrically active cells received its principal impetus from the remarkable work of Hodgkin and Huxley, in 1951, on nerve conduction in the squid giant axon. Hodgkin and Huxley used voltage-clamp methods to obtain extensive quantitative experimental results and proposed a system of ordinary differential equations that summarized and organized these data. Since then, their experimental methods have been extended and adapted to the study of other electrically active cells. Also, numerous mathematical studies of the Hodgkin–Huxley equations have been made. The results, experimental and mathematical, are scattered through the literature in research papers, and the first purpose of this book is to provide an organized account of some of these results. This account is intended to be accessible to mathematicians with little or no background in physiology.

In Chapter 2, a fairly detailed account is given of the experimental results of Hodgkin and Huxley, and similar detail is provided for the derivation of the Hodgkin–Huxley equations. It is not necessary to study the experimental results or the derivation of the equations in order to understand the equations themselves, which are a four-dimensional system of autonomous differential equations containing messy nonlinear functions. (The functions are, however, quite well-behaved: They are, indeed, real analytic functions, and the usual existence theorems can be applied to the differential equations.)

It is tempting to the mathematician to disregard the derivation of the equations and plunge ahead, instead, to the mathematical analysis of the equations, a familiar activity made additionally attractive, in this case, by the fact that the equations model an important system in the "real" world. There are, however, several reasons why this is an ill-advised procedure. First and perhaps foremost is that, without a knowledge of how the equations are derived, one has no understanding of the status of the various terms in the equations. There is a tendency to regard the equations as "written in stone" or as axioms, and to proceed from there. This viewpoint is often valuable in pure mathematics, but it is worse than useless in this part of applied mathematics. For this study, one needs to understand in what ways the equations are accurate descriptions and in what ways the equations represent rough descriptions of the physiological system. This kind of understanding can only be gained from some knowledge of how the equations are derived. Second, a knowledge of where the equations come from is essential in judging what mathematical problems to undertake in studying the equations. Without the guidance of that knowledge, the mathematician can easily drift into attractive mathematical problems that have no physiological significance. Finally, the great depth and ingenuity of the methods used by Hodgkin and Huxley in deriving their equations makes the study of the derivation a genuine intellectual adventure. It would be regrettable to miss it.

In the latter part of Chapter 2, we discuss some of the physiological phenomena that have been studied mathematically, and in Chapter 3 we describe some other models of nerve conduction, most of which were inspired by or are modifications of the Hodgkin–Huxley equations.

In Chapter 4, we describe some of the other models of electrically active cells that have been derived by modifications of and extensions of the methods of Hodgkin and Huxley. These descriptions are brief and incomplete. They are intended mainly to give the reader some idea of the models that exist and their status. In a later chapter, we will be particularly concerned with models of the cardiac Purkinje fiber.

The second objective of this book is to summarize the theory of ordinary differential equations that is used or may later be used in the study of these mathematical models. This summary (in Chapter 5) brings the mathematician to the research level in certain aspects of the theory and also indicates to the interested biologist the kinds of mathematics that may be useful in such studies. Our description of the mathematics is only a summary and omits all proofs and most examples.

We have a second purpose in Chapter 5: to emphasize a particular direction for study, that is, to emphasize the application of singular perturbation theory. The use of singular perturbation theory in nerve conduction models is certainly not new. Indeed, one of the first mathematical analyses of the Hodgkin–Huxley equations (actually a study of a simplified version of the Hodgkin–Huxley equations), made by FitzHugh in 1960, is a singular perturbation study in all but name. In the years since 1960, singular perturbation theory has been used in research papers, but it has not been used as systematically or as extensively as it could have been. Later in Chapter 6, we discuss explicitly the numerous reasons, both physiological and mathematical, for using the singular perturbation viewpoint. Our summary of the mathematics includes a detailed description of those results from singular perturbation theory that are clearly useful for the study of mathematical models of electrically active cells.

Our account of the mathematics to be used may seem disappointing for several reasons. To the physiologist, there may seem an undue emphasis on existence theorems and qualitative theory and not enough discussion of how to find explicit solutions. The answer to this may seem equally disappointing. Chapter 5 is intended as an account of the existing mathematics that seems to be useful in studying these mathematical models. The theory is by no means complete; much work of a purely mathematical nature remains to be done.

The next disappointment concerns the almost exclusive emphasis on ordinary differential equations. Partial differential equations play an important role in the study of nerve conduction. The most spectacular success of the mathematical work of Hodgkin and Huxley was the prediction of the speed of the nerve impulse, and that was obtained by analysis of a system of partial differential equations. However, it is by no means certain that the technique used by Hodgkin and Huxley to derive a system of partial differential equations is valid for other mathematical models of electrically active cells. The structure of other cells is more complicated than the cylindrical structure of the nerve axon, and this fact alone may preclude the derivation of similar partial differential equations for other cells. Second, as stated before, there are cogent reasons for making more extensive use of singular perturbation methods. If it turns out that the singular perturbation viewpoint is valuable in the study of the models that are ordinary differential equations, then this would suggest that singular perturbation techniques should also be used in the study of the partial differential equations. Thus, it seems reasonable to postpone study of the partial differential equations until more definite knowledge of how to approach the ordinary differential equations has been obtained.

In Chapter 6, we discuss the application of the methods, especially the singular perturbation theory described in Chapter 5, to the models of the electrically active cells derived in the earlier chapters. We give a detailed analysis of the FitzHugh–Nagumo equations, indicate the analysis that could be undertaken for the Hodgkin–Huxley equations, and discuss in some detail the analysis of the Noble model for the cardiac Purkinje fiber.

Although there is a logical order to the chapters in this book, certain sections are independent of one another. The reader with a serious interest in the subject can reasonably afford to omit parts of Chapters 3 and 4 on a first reading. How much of Chapter 5 needs to be read depends on the reader's mathematical background. (Remember that the singular perturbation theory described in Chapter 5 is applied in Chapter 6.)

A mathematician who is simply interested in seeing the applications of mathematics in nerve conduction might read Chapters 1 and 2 (despite the earlier stern injunctions, not every detail of the derivation of the Hodgkin–Huxley equations needs to be studied), the treatment of the FitzHugh–Nagumo equations in Chapter 3, the section in Chapter 5 on singular perturbation theory, and the discussion of the FitzHugh–Nagumo equations and the Hodgkin–Huxley equations in Chapter 6. A physiologist who is already acquainted with mathematical models of electrically active cells might omit most of the first four chapters and simply read Chapters 5 and 6.

In writing this book, I have received essential help from many people. I am especially indebted to Mark Kramer, who carried out the original calculations for Table 6.2, and to the students whose interest made possible the classes in which much of this material was developed. I am grateful to Dean John Yolton of Rutgers

College, who gave me the opportunity to present part of this work in undergraduate honors courses. Also, I wish to thank Carol Rusnak, who typed several versions of this book with unfailing good humor and patience.

I am happy to acknowledge support by the National Science Foundation Visiting Professorships for Women Program at the Courant Institute of New York University during 1984–85.

Finally, I am indebted to the staff at Cambridge University Press, especially Peter-John Leone, for thorough editing that did much to improve this book.

Jane Cronin

1

Introduction

The work of Hodgkin and Huxley on nerve conduction has long been recognized as an outstanding scientific achievement. Their papers were published in 1952 [Hodgkin and Huxley (1952a, b, c, d)] and they received a Nobel prize in physiology for their research in 1961. Hodgkin and Huxley's work was at once a triumphant culmination of many years of theoretical and experimental work by research physiologists and a pioneering effort that set the direction and defined the goals for much of the ensuing research in biophysics.

The purpose of this book is, first, to provide an introductory description of the work of Hodgkin and Huxley and the later work that is based on the techniques that they introduced. Our main emphasis is on the theoretical aspect of the Hodgkin–Huxley work, that is, the derivation and analysis of their mathematical models (nonlinear ordinary and partial differential equations); the second purpose of this book is to describe some of the mathematics that is used to study these differential equations.

The hope is that this book will indicate to some biologists the importance of the mathematical approach and will serve as an introduction for mathematicians to the mathematical problems in the field. However, this discussion is bound to be unsatisfactory to many readers. The biologists will find the description of the physiology simplistic, crude, if not actually misleading, and they may also be dubious about the value of conclusions that can be drawn from the mathematical analysis. Mathematicians who are accustomed to the precision and stability of physics and engineering, will find the inherent uncertainty of the parameters in the models dismaying, if not disagreeable. Also, despite some very successful analysis, the mathematical problems raised by these models remain largely unsolved.

Nevertheless, this discussion is useful because it emphasizes certain questions that must be resolved in the mathematical study of biological problems. The reply to the biologist who doubts the value of mathematical techniques is that those doubts may possibly be well founded. However, it is also true that mathematical results have shed significant light on certain questions in biology. The important activity should consist not in expressing doubts, but in advancing the study of the mathematical models to the point where it can be shown clearly that they are or are not an important aspect of biological study. To the mathematician who finds the problems difficult or unattractive esthetically, the reply is even simpler. These problems are here, and criticizing their origin or aesthetic value will not make them go away.

We shall assume that the reader is familiar with basic concepts from electricity, that is, potential, current, resistance, the units in which these are measured, and Ohm's law. Since capacitance is a somewhat less elementary electrical notion and because capacitance plays a very important role in the derivation of the Hodgkin–Huxley equations, a definition and elementary discussion of capacitance have been included in the short Appendix at the end of this book.

The problem of how a nerve impulse travels along an axon has a long and interesting history. A brief summary of this history and a number of references may be found in Scott (1975). Here we shall merely point out a couple of the results that made the work of Hodgkin and Huxley possible. The membrane that surrounds the axon had been discovered and its capacitance measured by Fricke (1923). Cole (1949) had pointed out that the important quantity to be measured was the potential difference across the membrane. Equally important was the discovery by Young (1936) of the squid giant axon. The unusually large diameter of this axon (about 0.5 mm) made experimental work possible.

The work of Hodgkin and Huxley, which was a study of how a nerve impulse travels along the squid giant axon, consisted of two parts. The first was the development and application of an experimental technique called the voltage-clamp method, which was invented by Cole. By using the voltage-clamp method, Hodgkin and Huxley obtained extensive quantitative data concerning the electrical properties and activities of the axon. The second part of their work consisted in deriving a mathematical model (a four-dimensional system of nonlinear ordinary differential equations) that summarized the quantitative experimental data. They then carried out a numerical analysis of the differential equations. As will be described later, that numerical analysis showed that the differential equations were remarkably successful in predicting a wide variety of experimental results.

The work of Hodgkin and Huxley had the additional importance that it set a direction for experimental and theoretical study of other electrically active cells, that is, cells whose electrical properties change during the normal functioning of the cells. Voltage-clamp methods have been developed for the study of myelinated nerve fiber (the squid axon has a particularly simple structure and is termed an unmyelinated axon), striated muscle fiber, and two kinds of cardiac fiber. For each of these, a mathematical model has been derived by using basically the same approach as that used by Hodgkin and Huxley in their study of the squid axon.

The mathematical analysis of the Hodgkin–Huxley equations and the analogous models for other electrically active cells consists of two different parts. The first part is numerical analysis, that is, computation of approximate solutions of the differential equations. Hodgkin and Huxley themselves carried out extensive numerical analysis in their original papers and obtained the most outstanding result of the theory: the prediction of the velocity of the nerve impulse. However, as we shall see in Chapter 2, the equations can be used to predict or describe many other experimental phenomena.

Many other numerical analyses of the Hodgkin–Huxley equations have since been made, and numerical analyses, have, until now, been the most useful results for physiologists. There are, however, two serious drawbacks to numerical analysis. First, although numerical analysis can yield much useful information (as in the example of the velocity of the nerve impulse), there are many important questions that cannot be approached by use of numerical analysis. Numerical analysis cannot yield an explanation of how the potential V and the sodium and potassium currents are related. As will later be shown, the simplest and crudest qualitative analysis yields far more information of this kind. Second, numerical analysis requires the assignment of strict numerical values to the parameters that occur in the differential equations. As we shall see later, the values of the parameters, indeed the very form of certain of the functions, are not known with much accuracy. Consequently, it is important to study mathematically some class or family of equations to which the Hodgkin–Huxley equations belong, as well as to study the equations themselves.

In Chapter 2 the work of Hodgkin and Huxley is described in some detail: first the experimental work and then the derivation of the equations. It is important to see a fairly detailed description of the experimental results and their interpretation even for a reader whose primary interest is the mathematical analysis of the equations. Only a knowledge of the origin of the equations makes clear the status of the equations and the significance to physiologists of various mathematical problems concerning the equations. Chapter 2 also summarizes some of the numerical analysis that was carried out by Hodgkin and Huxley. Some of this analysis was carried out on the equatons. However, Hodgkin and Huxley also derived from this original system a system of nonlinear partial differential equations (which we will term the full Hodgkin–Huxley equations) and they carried out a numerical analysis to find traveling wave solutions of the partial differential equations. It was this analysis that yielded the prediction of the velocity of the nerve impulse.

In Chapter 3 we describe some other mathematical models of nerve conduction including various simplifications and modifications of the Hodgkin–Huxley equations. Chapter 4 describes some mathematical models of other electrically active cells that were obtained by using the basic techniques and ideas introduced by Hodgkin and Huxley.

In Chapter 5 we turn to the problem of analyzing mathematically the models that have been described. This analysis requires two quite distinct kinds of mathematics. First, we need material from the subject of ordinary differential equations including the theory of singularly perturbed equations. This material is summarized in Chapter 5. In order to study the full Hodgkin–Huxley equations, considerable material from partial differential equations, in particular the theory of reaction-diffusion equations, is needed.

Rather than attempting to present this material, we have merely described it very briefly and cited a few references. There are several reasons for emphasizing the study of the ordinary differential equations and postponing a detailed study of the partial differential equations. First, the two kinds of theory are essentially independent of one another and represent two quite different subjects. Second, the ordinary differential equations are derived directly from the experimental data and hence are closer to the real world of physiology. [Very good results are obtained from studying the full Hodgkin–Huxley equations, but it is questionable whether the corresponding partial differential equations for other electrically active cells are realistic. See McAllister, Noble, and Tsien (1975), page 4.] Finally, it seems practical to deal with the ordinary differential equations in some detail first because this will help guide future work on the partial differential equations. For example, in Chapter 6 we shall discuss the reasons why it seems strategic to regard our models as singularly perturbed systems. Moreover, by using the singularly perturbed viewpoint we will obtain useful and enlightening information about how the electrically active cells behave. If more extensive research continues to show that the singularly perturbed viewpoint is valid and informative, then it will follow that the corresponding partial differential equations should be regarded as singularly perturbed systems. But that would suggest the use of very specific theory of singularly perturbed partial differential equations rather than general reaction-diffusion theory. Thus, to some extent, the study of the partial differential equations awaits the resolution of questions concerning the ordinary differential equations.

In Chapter 6 we use the theory from Chapter 5 to study the models derived earlier. In particular, we make a detailed study of the Noble model of the cardiac Purkinje fiber. Also we summarize very briefly some of the work on traveling waves in nerve conduction, that is, traveling wave solutions of the full Hodgkin–Huxley equations.

2

Nerve conduction: The work of Hodgkin and Huxley

2.1 The physiological problem

We start by describing in some detail the physiological problem to be studied. An impulse that, say, carries a command from the brain to a muscle travels along a sequence of neurons that can be portrayed roughly as in the sketch in Fig. 2.1. When the impulse arrives at the dendrites on the left-hand side of the neuron, the stimuli given the dendrites are integrated at the cell body to form a nerve impulse. The nerve impulse travels along the axon to the branches of axons on the right-hand side of the neuron. The impulse then jumps to another set of dendrites and the process just described is repeated. Neurons vary considerably in size. The sciatic nerve of the giraffe contains an axon that may be several meters in length; many other axons are much shorter. The diameter of the squid axon, which has been the subject of many experimental studies, is ~ 0.5 mm but it can be as much as 1 mm. Its length is several centimeters.

The process by which an impulse travels along a sequence of neurons is quite complicated. For example, when the impulse jumps from one set of dendrites to another, both chemical and electrical processes, which are not well understood even today, play an important role. The subject of Hodgkin and Huxley's work is the process by which the impulse travels along the axon in the giant axon of the squid. The reason that Hodgkin and Huxley studied the squid axon is that its diameter is unusually large and consequently makes experimental work possible. (The discovery of the squid giant axon by Young in 1936 was an important step in the development of cell electrophysiology because it made a whole new direction of experimentation possible. We note that the term "giant" axon is a relative one: It refers simply to the fact that the diameter

SKETCH OF NEURON

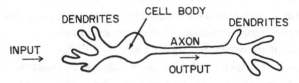

Figure 2.1.

of the axon is generally ~ 0.5 mm, a much larger diameter than that of most axons.)

Restricting the investigation to the question of how an impulse travels along an axon may seem to be a very narrow problem. In actuality, it is a wide and important problem. As we shall see later, the processes by which an impulse travels along an axon are complicated and still not very well understood; thus the problem is certainly deep. The techniques introduced by Hodgkin and Huxley, both experimental and mathematical, have been adapted and extended so that they can be used to study other quite different electrically active cells. Among these are myelinated axons, striated muscle fibers, and cardiac fibers. All of these will be described later. Finally, in the study of the squid axon, the membrane covering the axon plays a very important role in the electrical processes. Any progress made in the study of how the nerve impulse travels along the axon is necessarily progress in the study of membrane biophysics. Thus the work of Hodgkin and Huxley is both broad and profound.

In this work the axon will be regarded as having the shape of a cylinder (this is a significant simplification). The axon consists of homogeneous aqueous matter called axoplasm, and the axoplasm is surrounded by a thin membrane (< 100 Å thick) [Å (angstrom) = $m/10^{10} = mm/10^7$]. The membrane is composed of two types of macromolecules: protein and lipid (fat). We assume that the squid axon has this simple uniform structure: a cylinder of axoplasm surrounded by a thin membrane of uniform thickness. However, in vertebrates, except for the smallest axons, the structure is more complicated. The axon is myelinated, that is, the axon is surrounded by a sheath of fatty material called myelin. This sheath of myelin is interrupted at intervals of about 1 mm by short gaps called nodes of Ranvier. In these short gaps, the axon consists of axoplasm surrounded by the thin membrane previously described. The simpler axon structure, of which the squid axon is an example, is called an unmyelinated axon. Later, it will be shown that the methods of Hodgkin and Huxley, which were developed for the study of an unmyelinated axon (i.e., the squid axon), can be adapted to the study of a myelinated axon.

Although we will not be concerned further with the subject until some time later, we point out here that the speed with which a nerve impulse travels along the axon depends on the diameter of the axon and on whether the axon is myelinated. If the axon is unmyelinated then as the diameter of the axon increases, the velocity of the impulse increases. Indeed, the giant axon of the squid transmits "escape signals," and thus there is good reason for rapid transmission and hence for an axon of large diameter. But there are practical limitations on the diameter of the axon; in vertebrates, the fast transmission of nerve impulses has been obtained by a different evolutionary development. It turns out that if the axon is myelinated, the transmission of the nerve impulse is much faster even if the diameter of the axon remains small. For a quantitative description and analysis of how the speed of the nerve impulse depends on the diameter of the axon and whether the axon is myelinated, see Chapter 4 and Scott (1975, p. 516).

Now we give a brief description of some of the history of this problem so that we can indicate the status of the problem when Hodgkin and Huxley began to work on it. Experimental studies of this physiological problem had been carried on for a long time. By the end of the eighteenth century, it had been established that nerves could be stimulated by an electric shock and in 1850, Helmholtz measured the signal velocity on a frog's sciatic nerve. This work suggested that the basic problem of nerve conduction is electrical in nature and one might conjecture that the axon is a fairly effective biological substitute for an everyday conductor of electricity like copper or silver wire. That is, one might conjecture that nerve conduction consists of an electric current. There are two cogent arguments against this idea. First, the Helmholtz experiment in 1850 showed that the signal velocity is fairly slow (27 m/s) compared to the speed of electric current in a wire, which is the speed of light. Second, although the interior of the nerve fiber contains ions and is a fairly good conductor of electricity, the diameter of the fiber is so small that the resistance per unit length of a nerve fiber of average diameter is $\sim 10^{10}$ Ω/cm. This is an enormous resistance. It means that the electrical resistance of a nerve fiber of length 1 m is about that of 10^{10} miles of 22 gauge copper wire. (Copper wire, gauge 22, has a diameter of 0.7112 mm and 10^{10} miles is about 10 times the distance between Earth and the planet Saturn.) By Ohm's law, we have

$$(\text{potential}) = (\text{current})(\text{resistance}).$$

If a nonnegligible current were to be produced with such a large resistance, a huge potential would be necessary. Because such potentials certainly do not occur in animals, this suggests that the nerve impulse is not carried directly by an electric current but is in some way reinforced as it travels along the nerve fiber. One of our objectives is to find out how this process occurs.

The objective of the work of Hodgkin and Huxley was to explain not only the reinforcing process, but also the many phenomena observed in experiments conducted by physiologists over a long period of time. Summarizing the salient features of these experiments is our next step. When the giant axon is in the living squid, there is a potential difference across the membrane surface, and this potential difference remains practically unchanged if the axon is carefully removed from the squid. The isolated axon "survives" for many hours after removal from the body of the squid in the sense that its electrical properties remain about the same. As a result, it is possible to study the isolated axon experimentally. The crucial experimental observations concern how the potential difference across the membrane, called the membrane potential, changes if the nerve is stimulated, and our summary consists essentially in describing the types of changes in membrane potential that occur.

If the axon is not subject to stimulation, the inside of the axon is 50 to 70 mV (millivolts) less than the outside. Today the usual sign convention is to denote this by -50 to -70 mV (i.e., inside minus outside). Hence we say that if the axon is at rest (not subject to stimulus) the membrane potential is -50 to -70 mV. This value for the membrane potential is called the *resting potential*. As long as the axon is not stimulated, the resting potential remains fairly constant. In this chapter, we will regard it as a strict constant. Later, we shall see that a more realistic view requires the introduction of an element of randomness in the description of the resting potential. Clearly a question of basic importance is that of how the resting potential is maintained. We will deal with this question a little later.

Now suppose that a stimulus in the form of a brief current pulse consisting of a flow of positive ions and < 1 ms in duration is applied to the axon by an electrode inserted into the interior of the axon. The resulting change in the membrane potential depends on the amplitude of the stimulus in the following way.

If the stimulus is very weak, there is a temporary positive change in the membrane potential that is proportional to the amplitude of the stimulus, which dies away in ~ 1 ms and which affects the membrane potential only very near the point where the membrane is touched by the electrode providing the stimulus. The response is said to be passive and local. If the amplitude of the stimulus is increased, the membrane potential begins to increase faster than the stimulus amplitude (i.e., it is no longer simply proportional to the stimulus amplitude), but the change in membrane potential does not exceed $+10$ mV. The response is said to be active and local.

If the amplitude of the stimulus is large enough so that the membrane potential is raised above a critical value called the *threshold*, ~ -40 mV, then the membrane potential increases abruptly to form a roughly triangular solitary wave $\sim +30$ mV in amplitude and lasting ~ 1 ms. (This interval depends on the temperature.) This curve is called an *action potential*. Although it arises locally, it splits immediately into two separate waves that travel away from the stimulating electrode in opposite directions along the fiber at a constant conduction velocity. The action potential, once it starts traveling, can be recorded along the nerve fiber as it passes, but there is no way to deduce from the record alone where the action potential originated or the amplitude of the stimulus that produced it. The fact that only the presence or absence of the traveling wave can be recorded is called the all-or-nothing law in physiology. [It is interesting to note that an all-or-none mode of conduction is an efficient way to avoid certain interference and disturbance; see Hodgkin (1971, p. 16).]

Following a first stimulus that produces an action potential there is a time interval of about 1 ms called the absolute refractory period during which no stimulus, however strong, can produce an action potential. After the absolute refractory period there is a relative refractory period during which an action potential can be evoked but only by a stimulus of larger amplitude than the usual threshold value. During the relative refractory, the required larger amplitude gradually decreases to the usual threshold value. As we shall soon see, the refractory period plays a crucially important role in the propagation of the action potential.

If the amplitude of the first stimulus is below the threshold value, no action potential is produced, but response to a second stimulus is affected for an interval afterward. First there is an enhanced phase ~ 1–2 ms during which the threshold is lower than the usual value. Thus the effects of two stimuli, applied close enough together in time, combine, and the closer together they are, the more effective. After the enhanced phase, there is a depressed phase of indefinite duration during which the threshold to the second stimulus is greater than the usual threshold.

If the stimulus is a very long current pulse (say 0.5 s), more than one action potential may result. Depending on the amplitude of the stimulus, there may be as many as four or five action potentials in an interval as short as 100 ms. (Such a group of action potentials is sometimes called a burst.) The action potential amplitudes decrease somewhat during the burst.

The results that have just been described are obtained from many difficult and delicate experiments. In order to measure the membrane potential, a microelectrode must be introduced into the nerve without disturbing or "killing" it, and then very small potentials must be measured. The preceding brief description roughly summarizes a huge amount of hard work. [For an indication of some of the difficulties in this work, see the amusing description given by FitzHugh (1969, p. 70).]

2.2   A brief summary of Hodgkin and Huxley's conclusions

The purpose of the work of Hodgkin and Huxley was to obtain a physical explanation of the processes that produce the experimentally observed phenomena that have been described. Their results are a sharply quantitative description of currents that flow across the membrane surrounding the axon and an explanation of how these currents produce the experimental results previously described. Before starting a detailed description of the work of Hodgkin and Huxley we will give a crude brief summary in words of the explanation proposed by Hodgkin and Huxley. This loosely written summary will serve as a guide in the later more detailed account of Hodgkin and Huxley's work.

Hodgkin and Huxley's explanation is largely based on a description of varying currents of sodium and potassium ions across the membrane that surrounds the axon, and our brief summary will be concerned exclusively with these currents.

Suppose first that the nerve fiber is in its natural environment, that is, intercellular fluid that has a concentration of sodium (Na) ions and chloride (Cl) ions about equal to that of sea water, and suppose that the nerve fiber is not subject to any stimulus. It is known that the concentration of Na ions outside the axon is about 10 times the concentration of Na ions inside the axon and that the concentration of potassium (K) ions inside the axon is about 5 times the concentration of K ions outside the axon. The difference in concentration across the membrane produces a chemical driving force on the ion that is equivalent to an electromotive force or potential difference equal to

$$\frac{RT}{F} \ln \frac{C_i}{C_o}. \tag{2.1}$$

C_i and C_o are the ionic concentrations inside and outside the axon, respectively, R is the gas constant, which has the value 8.31432×10^7 erg/°C (i.e., degree Centigrade) per mole, T is the absolute temperature, and F is the Faraday number which is the product of Avogadro's number and the charge on one electron [i.e., $F = 96,500$ C (coulombs)]. Expression (2.1) is called the Nernst formula. For a discussion of its derivation, see the article by Schwartz in Adelman

(1971). Thus, the electromotive force acting on the Na ions is

$$\frac{RT}{F}\ln(10^{-1}) = -\frac{RT}{F}\ln 10 = -55 \text{ mV}$$

and the electromotive force acting on the K ions is

$$\frac{RT}{F}\ln 5 = 75 \text{ mV}.$$

The signs of these electromotive forces are chosen so as to be consistent with the convention chosen earlier in the description of the resting potential. According to that convention, a positive electromotive force (emf) is an emf that "pushes" positive ions (electrons) in the direction across the membrane from the inside (outside) toward the outside (inside) of the axon. Similarly, a negative emf "pushes" positive ions from the outside toward the inside of the axon.

Besides the chemical driving force on the Na and K ions, there is also the force of the resting potential, which, as mentioned earlier, has the value -70 mV. We postpone the question of how the resting potential is maintained until after further discussion of the currents of Na and K ions. For the present we concentrate on the question of how ionic currents are produced by the resting potential and the emfs that act on Na and K ions.

First, the emf equivalent to the chemical driving force acting on K ions is 75 mV, which is slightly larger than the absolute value of the resting potential and is opposite in sign from the resting potential. Hence if the membrane is permeable to K ions, i.e., allows K ions to pass through it, then we would expect that there would be either no flow or a small outward flow of K ions. This is, in fact, what occurs. On the other hand, the emf acting on the Na ions is -55 mV, which has the same sign as the resting potential. Thus the total emf acting to push Na ions across the membrane from the outside toward the inside of the axon is

$$-55 - 70 = -125 \text{ mV}.$$

Hence we would expect (if the membrane were permeable to Na ions) that when the nerve fiber is in its natural environment and not subject to stimulus, there would be a significant inward current of Na ions. Now, in fact, there is no such inward flow of Na ions; so we conclude that the membrane is impermeable to Na ions under the conditions described. It turns out that the permeability of the membrane to Na and K ions is dependent on the potential difference across the membrane. As we shall now see, this fact is of crucial importance in explaining how an impulse travels along the nerve axon. That is, we will now describe the ionic currents that take place when some stimulus is applied to the axon so that the potential difference across the membrane is changed.

If the membrane potential is made more negative (this is called hyperpolarization), the membrane permeability to ionic currents remains unchanged. Hence, there is only a very small outward flow of K ions (or none) and no flow of Na ions. However, if the membrane potential is increased (this is called depolarization) to a certain critical potential, the threshold potential (about -40 mV), the permeability of the membrane to Na ions increases, Na ions enter the nerve fiber at an increasing rate, and the membrane potential moves rapidly toward the equilibrium potential of the Na ion, that is, the potential that just balances the equivalent emf produced by the chemical driving force acting on the Na ions. As our earlier discussion showed, this value is 55 mV. Depolarization also causes an increase in the conductance of the membrane to K ions, although this increase is somewhat slower than the increase of conductance of Na ions. The threshold is the membrane potential at which the inward Na ion current just balances the outward K current.

For a subthreshold depolarization, the outward K current is larger than the inward Na current and the membrane potential returns to the resting potential. For a superthreshold depolarization, the inward Na current exceeds the outward K current and continues to increase until the membrane potential reaches the equilibrium potential of the Na. As the equilibrium potential of the Na ion is reached, the permeability of the membrane to Na ions decreases fairly rapidly, but the permeability to K decreases much more slowly. Since the K current is an outward ionic current, the membrane potential then returns to the resting potential. (It should be emphasized that the description that has just been given is very rough. Neither the magnitude nor the timing of the events are indicated precisely. Later we will consider experimental data and the mathematical model that summarize the data. These will provide us with a more precise description.) Meanwhile, the action potential is propagated along the nerve fiber because points on the fiber near the point under observation have their membrane potentials depolarized to the threshold value. It is important to observe the crucial role that the refractory period plays at this stage. If the action potential has reached a certain point, then nearby points to which the action potential has not yet arrived (say points on the right in the accompanying diagram) will be depolarized to threshold value. But nearby points through which the action potential has just passed will be in the refractory period and hence will not be depolarized to threshold value. Thus the action potential is propagated unilaterally, which is the desired objective.

$\rightarrow$ direction of propagation

| nearby points in | nearby points not in |
| refractory period | refractory period |

The net result of the formation of the action potential is that the nerve fiber gains some Na ions and loses some K ions. In the formation of a single action potential, these gains and losses are very small. However, they are cumulative and it is natural to look for some process that serves to offset them. Physiological studies suggest that there are mechanisms driven by metabolic energy that pump out Na ions and pump in K ions. For a brief description of these pumps, see Luria (1975, pp. 334–335). However, there are experimental anomalies that cannot be explained by such mechanisms [see Beall (1981, p. 8)].

Now we return to the question of how the resting potential is maintained. When the membrane is at rest (i.e., not subject to any stimulus), it is highly permeable to K ions, that is, K ions flow through the membrane easily, but the membrane is only slightly permeable to Na ions. Because the concentration of K is much higher inside the axon than outside and because the K ions flow easily through the membrane, there is a leakage of K ions from the inside to the outside of the axon. Thus, the inside of the axon becomes electrically negative with respect to the outside. That is, an electrical field is set up that tends to push positively charged particles from the outside to the inside. The outward flow of K ions continues until this electrical field becomes strong enough to balance the chemical driving force that acts on the K potassium ions and that is equivalent to the electromotive force given by the Nernst formula. The potential to which the electric field gives rise is the resting potential.

2.3 The work of Hodgkin and Huxley
2.3.1 The experimental results

So far we have stated the physiological problem that Hodgkin and Huxley worked on and we have described briefly and loosely the qualitative aspect of Hodgkin and Huxley's solution of the problem. Now we must go back and describe in detail how Hodgkin and Huxley arrived at their answer to the problem. If we compare the study of nerve conduction to a murder mystery, we might say that we have presented the first and last chapters of the story. We know that "the butler did it", but now we must go through the painstaking reasoning by which Hercule Poirot arrived at the identity of the murderer. We want also to obtain the quantitative version of the answer. That is, we want to replace statements such as, "this increase [of conductance of potassium ions] is somewhat slower than the increase of conductance of sodium ions," with precise quantitative statements.

Our analysis consists of two parts. First we describe the experimental results and the analysis of these results. Next, on the basis of these results, we derive the Hodgkin–Huxley differential equations that can be regarded as a quantitative summary of the experimental data.

The loose description of Hodgkin and Huxley's conclusions that we have already given suggests the nature of the experiments that they carried out. It seems reasonable to assume that Hodgkin and Huxley decided somehow that Na ion current and K ion current played an important role in the formation of the action potential. Then they measured the time-varying flows of Na and K current at a fixed point on the surface of the axon as an action potential passed by. In theory, this might be a good approach to use. In fact, it is not possible to develop techniques for directly measuring the flows of different ionic currents. Consequently, indirect approaches must be used and often considerable ingenuity and effort is required in order to interpret the experimental data. Moreover, in order to interpret the data, it is frequently necessary to make certain assumptions. These assumptions are justified later in the sense that the conclusions based on them agree with experimental results and can be used to predict other results. (We will presently see examples of these assumptions and their justifications.)

It is crucially important to have a clear picture of this structure of experimental data, interpretation of the data, and the assumptions upon which the interpretation rests because this structure is the foundation of the mathematical model and is a measure of the status of the model. A good judgment of what is a sensible mathematical problem upon which to work must be based in part on a understanding and knowledge of the status of the mathematical model (the Hodgkin–Huxley equations). It is for this reason that we will describe the experimental results in more detail than may at first seem necessary in a discussion of the mathematics of nerve conduction.

We will describe in some detail the experiments in the first two papers in the series on nerve conduction, that is, Hodgkin, Huxley, and Katz (1952) and Hodgkin and Huxley (1952a). In the paper by Hodgkin, Huxley, and Katz (hereafter to be referred to as HHK) there are two sets of experiments. The first experiments are those in which a brief stimulus is applied to the nerve. (Experiments of this kind have already been discussed Section 2.1.) Experiments of this type had been performed in earlier work, but it was necessary to carry them out at the start of the study in order to establish that the axons studied were capable of giving action potentials [see HHK (1952, p. 433)].

Although we will describe the results of the experiments in some detail, we will indicate the experimental procedure only briefly. There are several reasons for this omission. Most important is the fact that it is not immediately relevant to our goal. Second, the electrical circuitry used is fairly complicated [see HHK (1952, p. 430) for a diagram of the feedback amplifier used to maintain the membrane potential at a fixed value]. Finally, this circuitry is now completely outdated (radio tubes, i.e., "valves" in the terminology used by Hodgkin and Huxley, have long since been replaced in electrical circuits by transistors). The brief treatment of experimental procedure is misleading in one important respect. The reader may fail to realize the hardships and complications that are encountered in carrying out these experiments. It is worthwhile to read the description of the experimental procedure given by HHK (1952, pp. 426–432) just to get an idea of the pitfalls and difficulties that beset the experimenter.

Before proceeding to the description of the experiments, we make one remark concerning quantitative biological data. The "constants" that occur in biological data are not constant values in the same sense as constants in physics, such as the gravitational constant. For example, we have already introduced the concept of resting potential and used as its value -70 mV. In fact, the value of the resting potential varies from axon to axon and is also dependent on the condition of the axon. Indeed, as we will see later, the resting potential is more accurately represented by inclusion of a stochastic term and this fact may be important in certain analyses. The point to be emphasized here is that measurements vary from experiment to experiment and, moreover, we must expect this to occur.

In the first set of experiments described in HHK, a fine silver wire electrode is inserted into the axon, a brief pulse of current or shock is applied to the electrode, and the changes in the membrane potential are recorded. (Remember that the preceding sentence is a simplistic description of the experiment!) Note that in this experiment, the quantities considered are uniform along the length of the axon. That is, the charge is uniformly distributed along the electrode and the potential difference across the membrane is the same at all points along the axon. Such an experiment is sometimes called a space-clamp experiment.

Note also that the events that occur during a space-clamp experiment are artificial. That is, such events do not occur in the normal functioning of the axon. In a space-clamp experiment, the membrane potential has, at any given time, the same value of all points on the axon. As time elapses, the membrane potential changes, but it is constant along the axon. More precisely, the membrane potential is a function of time but is independent of position on the axon. By contrast, in the normal functioning of the axon, the membrane potential has the same value, the resting value, all along the axon when the axon is not subject to any stimulus, but if the axon is stimulated, say by a pulse of current, then the membrane potential changes significantly near the point where the stimulus is applied. If the stimulus is large enough, an action potential travels along the axon. In these circumstances, the membrane potential at a given time, regarded as a function of position along the axon, is certainly not a constant function.

The magnitude and duration of current pulse in the experiment just described are measured and consequently the total electric charge in the shock can be measured. Measurement of the membrane potential shows that the initial change in the membrane potential is proportional to the total electric charge in the shock. The proportionality constant thus computed is the capacitance per unit area of the membrane. This capacitance, which has the value of ~ 0.9 $\mu F/cm^2$, plays an important role in the later analysis as we shall see.

The changes that take place in the membrane potential following application of a shock are indicated for some typical cases in Fig. 2.2, where the membrane potential is graphed against time. The graphs obtained depend both quantitatively and qualitatively on the magnitude and sign of the shock.

Let us consider first the case in which the shock consists of a surge of electrons into the electrode. Then the membrane potential, measured in accordance with the previously described convention first decreases (this is called *hyperpolarization*) and then as time passes, increases monotonically to the resting value. Only two graphs of this kind are sketched (graphs of membrane potentials that were reduced from the resting value by -5 and -39 mV). However, the behavior of the membrane potential in these two cases is typical of experiments in which hyperpolarization occurs. Now consider the case in which the shock consists of a surge of electrons away from the electrode (a surge of positively charged particles into the electrode). Then the membrane potential increases (this is called *depolarization*) and the graph of the membrane potential as a function of time depends very strongly on the magnitude of the shock. If the depolarization is less than 15 mV,

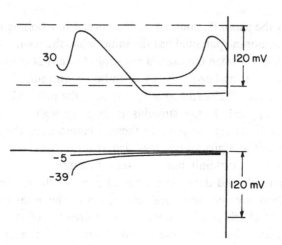

Figure 2.2. Time course of the membrane potential following
a short shock at 6°C. Depolarization shown upward. Numbers
attached to curves give the strength of the shock in
$m\mu$ coulomb/cm^2.

the membrane potential returns monotonically to the resting potential, that is, there is a subthreshold response. (No graph of this is
shown in Fig. 2.2) If the depolarization is more than 12 or 15 mV,
then, as the graph suggests, an action potential occurs. Notice that
if the depolarization is between 12 and 15 mV, then a state of
unstable equilibrium occurs and the graph shows that it may last
for some time before an action potential is produced or there is a
subthreshold response.

It should be pointed out at this stage that the sign convention
used by Hodgkin and Huxley in measuring the membrane potential
is exactly the opposite in sign from the convention used today and
previously described. Consequently in Hodgkin and Huxley's convention the membrane potential increases with hyperpolarization
and decreases with depolarization. However, the graphs in Fig. 2.2,
which are taken from HHK, represent depolarization as an increase
and hyperpolarization as a decrease. That is, for these graphs,
HHK departed from their own convention. In doing so, they used a
convention that coincides with the standard convention today.

The shocks that are applied in the experiments just described
have duration of about 8 μs. It is observed experimentally that the

total current I becomes negligible (can be considered zero) following an interval of 200 μs after the shock. Consequently, at that time, the equation

$$I = C_M \frac{\partial V}{\partial t} + I_i$$

(see Appendix) becomes

$$I_i = - C_M \frac{\partial V}{\partial t}.$$

We know the value of C_M and if a fixed time $t > 200$ μs is chosen, the value of dV/dt can be measured geometrically on the earlier graphs (Fig. 2.2). The value dV/dt is simply the slope of the graph at $t = t_0$. Thus the ionic current I_i can be computed at a fixed time $t > 200$ μs. In Fig. 2.3, I_i at 290 μs is plotted against the initial displacement of the membrane potential. Following the convention used by HHK, we regard the initial displacement V to be positive if hyperpolarization occurs and regard V as negative if depolarization occurs. The resulting graph in Fig. 2.3 summarizes a lot of experimental data and gives us insight into the relation between current and initial displacement of the membrane potential. It also affords us our first example of how further information can be extracted from the explicitly given experimental data.

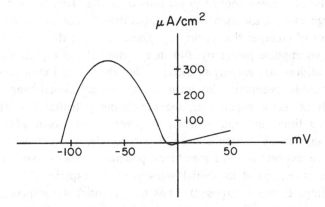

Figure 2.3. Graph of ionic current density against displacement of membrane potential.

The next experiments are the first voltage-clamp experiments. A *voltage clamp* is a sudden displacement of the membrane potential from its resting value to a new level at which it is held constant by electronic feedback. In the voltage-clamp experiments, a voltage clamp is established and the membrane current is measured during a subsequent interval of time. Voltage-clamp experiments were made possible by the earlier work of a number of physiologists, especially Cole (1949), who introduced the technique. The work of Hodgkin and Huxley consists of a painstaking and brilliant development of this experimental technique and an original, highly successful theoretical analysis of the resulting experimental data. A measure of the power and influence of their accomplishments is the fact that since their work, most experimental and theoretical studies of electrically excitable cells have been based on the techniques they developed.

We will describe the results of a series of voltage-clamp experiments in which various values are taken for the voltage clamp (the displacement of the membrane potential) and in which the axon is immersed in fluids with various Na ion concentrations. The resulting experimental data, when analyzed with sufficient care, yield a tremendous amount of information. However, it is important to make clear that a voltage-clamp experiment consists of imposing entirely artificial conditions on the axon. In the earlier experiments (space-clamp experiments) described, the axon is stimulated by a pulse of current applied to the interior of the axon. If the pulse is large enough, the changes in potential that then occur are the same kind of changes that occur at a given point on the axon when a nerve impulse passes by. But in a voltage-clamp experiment, the conditions are entirely artificial in that they do not even remotely resemble events in the axon in its normal functioning. In a voltage-clamp experiment, the membrane potential has, at any given time, the same value at all points on the axon. (Thus the voltage-clamp experiment is also a space-clamp experiment.) But, more importantly, the membrane potential remains fixed as time passes (except at the initial moment of the experiment when the voltage clamp is exposed). Thus the potential is independent of position along the axon and independent of time. In the normal condition of the axon, the membrane potential has the same value, the resting value, all along the axon when the axon is not subject to any stimulus. But if the axon is stimulated, say by a pulse of current, then the membrane potential changes significantly near the point where the stimulus is applied. If the stimulus is large enough, an action potential travels along the axon. In these circumstances, the membrane potential varies both with position along the axon and with time.

Let E denote the membrane potential and E_r the resting potential, and let $V = E - E_r$. The main reason that voltage-clamp experiments yield so much useful information lies in the basic electrical equation described in the Appendix:

$$I = C_M \frac{\partial V}{\partial t} + I_i.$$

In a voltage-clamp experiment, there is a sudden displacement or change of potential, that is, V is changed suddenly, and then V is held constant. Hence there is initially a surge of capacitance current, contributed by the term $C_M(\partial V/\partial t)$, but after that initial surge, V is constant and, hence, $\partial V/\partial t$ is zero. Hence the capacitance current is zero and so the total I across the membrane consists of ionic current I_i, that is,

$$I = I_i.$$

But I can be measured and hence in a voltage-clamp experiment, the ionic current I_i can be measured.

Now we describe a few typical results of voltage-clamp experiments performed on axons immersed in sea water, which is approximately the natural environment of the axon.

If the membrane potential is decreased from the resting value -70 to -135 mV (i.e., if V is -65 mV), then, as shown in Fig. 2.4, there is a brief surge of capacitance current followed by a fairly constant inward current (of positive ions) of about 30 $\mu A/cm^2$.

If V has any other negative value or if V has a positive value < 10 mV, similar results are obtained; there is a brief surge of capacitance current followed by a constant (with time) current I of positive ions. If V is negative, the current is inward; if V is positive, the current is outward.

If V is negative, there is a linear relation between V and I. Thus the results if V is negative resemble what would be expected from

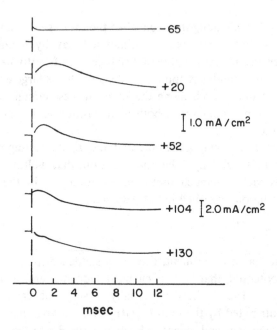

Figure 2.4. Record of membrane current under voltage clamp. Displacement of the membrane potential in mV is given by the number attached to each record. [The potential values are given in terms of the present conventions, i.e., inside minus outside, which is the opposite of the convention used by Hodgkin, Huxley, and Katz (1952).] Inward current is upward deflection.

considerations in physics if the membrane were regarded as a conductor with constant resistance per unit area, that is, we obtain a kind of Ohm's law. (This is, of course, a rough statement because in Ohm's law, V is the potential difference whereas here V is the deviation of the membrane potential from the resting potential and we assume that, for whatever reason, there is no significant flow of current across the membrane when the membrane potential is equal to the resting value.)

However, if V is positive and ≥ 10 mV, then the ionic current has very different kind of behavior. This is indicated in Fig. 2.4. Unlike the results when V is negative, a version of Ohm's law no longer provides any guidance. When V is positive and > 10 mV the initial flow of ionic current (after the surge of capacitance current) is an inward flow of positive ions. But after a short interval of time, the flow of ionic current reverses its direction.

Because all this time the membrane potential is constant, it is clear that, in these conditions, we cannot think of the membrane as having a constant resistance. Indeed as will be gradually revealed there is a complicated quantitative relationship among the ionic currents and V. (None of this is very surprising in view of our earlier discussion in Section 2.1 because if $V > 10$ mV, the membrane potential is approximately equal to or greater than the threshold value. Consequently, we expect nonlinear relationships to occur.)

We emphasize once again that our descriptions of the experiments are simplistic. Moreover, besides omitting technical details, we are also omitting certain of the scientific results. For example, all the results we have described are affected by temperature and a complete quantitative account requires taking the temperature into account. Because our purpose is primarily to indicate the steps taken to obtain the mathematical description rather than to give a fully detailed account, we will omit all references to temperature.

In the previous experiments, the ionic current under various voltage clamps has been studied. The next step is to identify the various ions that make up the ionic current. There was earlier evidence [see Hodgkin (1951)] to indicate that the initial "hump" of inward ionic current that occurs if V has value like 20 mV is composed largely of sodium. In order to investigate this, Hodgkin and Huxley (1952a) conducted voltage-clamp experiments similar to the ones already described in which the axon was immersed in sea water. The only difference was that the axon was immersed in various fluids with a lower concentration of Na ions than the concentration in sea water. From the results of these experiments, fairly direct inferences can be made concerning the presence or absence of Na ions in the ionic current. These experiments are described in Hodgkin and Huxley (1952a), which is the most outstanding of the experimental papers both in originality and accomplishment. The voltage-clamp experiments described in this paper were performed on:

(i) Axons immersed in sea water.
(ii) Axons immersed in choline sea water (i.e., sea water in which the Na ions have been replaced by choline). Choline is an inert biochemical ion with properties such that if Na

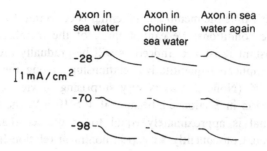

Figure 2.5. Membrane current during voltage clamps when the axon is immersed in sea water and in choline sea water.

ions are completely replaced by choline, then the squid axon becomes completely inexcitable, but the resting potential is about the same as if the axon were in sea water.

(iii) Axons immersed in 30% Na sea water (i.e., sea water in which 70% of the Na ions have been replaced by choline).

Typical experimental results are indicated in Figs. 2.5 and 2.6. The

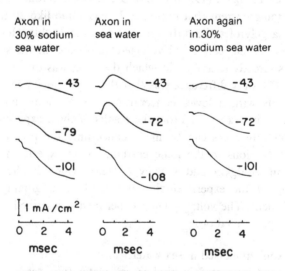

Figure 2.6. Membrane current during voltage clamps when the axon is immersed in sea water and in 30% sodium sea water.

most important features shown in Fig. 2.5 are the following:

1. If the external Na concentration is zero, that is, if the axon is immersed in choline sea water, the initial inward current is zero and there is an early increase in the outward current.
2. When the axon is immersed in choline sea water, the later outward current is only slightly altered. It is ~ 15–20% less than the later outward current that occurs if the axon is immersed in sea water.

With different values for the voltage clamp, similar records are obtained. These results give qualitative support to the hypothesis that the early inward current is carried by Na ions. These experimental results can also be used to give quantitative support to this hypothesis. To show this, we introduce the notion of sodium potential. According to the Nernst formula, the chemical driving force pushing the Na ions inward is equivalent to the emf equal to

$$\frac{RT}{F}\ln\frac{[\text{Na}]_i}{[\text{Na}]_o},$$

where $[\text{Na}]_i$ and $[\text{Na}]_o$ are the Na ion concentrations in the interior and the exterior of the axon, respectively. If the membrane potential equals

$$-\frac{RT}{F}\ln\frac{[\text{Na}]_i}{[\text{Na}]_o} = \frac{RT}{F}\ln\frac{[\text{Na}]_o}{[\text{Na}]_i},$$

then the emfs just balance one another and so the net flow of Na ions across the membrane is zero. The number (RT/F) $\ln([\text{Na}]_o/[\text{Na}]_i)$ is called the *sodium equilibrium potential* and is denoted by E_{Na}. The value of the sodium potential when the axon is immersed in sea water or 30% sea water can be estimated by examining the graphs that are the result of voltage-clamp experiments. If we assume that the initial ionic current flow is a flow of Na ions (we have already seen qualitative evidence to support this hypothesis), then if the value V of the voltage clamp is such that there is no initial ionic flow, then $V + E_r$, where E_r is the resting potential, must be the sodium potential. Inspection of Figure 2.6 shows that if the axon is in 30% sea water, the sodium potential is very close to 79 mV, and if the axon is in sea water, the sodium potential is slightly less than 108 mV. *That is, the voltage-clamp experiments yield fairly accurate estimates of the sodium potential.* Now we use these estimates of the sodium potential to obtain quantitative support for the hypothesis that the initial ionic current flow in a voltage-clamp experiment is a flow of Na ions.

To do this, the reasonable procedure would seem to be to compute E_{Na} by using the Nernst formula and to compare the result with the observed experimental value (close to 79 mV) just described. The difficulty in such a direct procedure is that the value of $[Na]_i$ is needed, but $[Na]_i$ is not known with any accuracy for the axon being studied experimentally. (Indeed, to measure $[Na]_i$ would require destruction of the axon.) Consequently, the following more indirect procedure is used.

Let $[Na]_o$ and $[Na]'_o$ denote the Na ion concentration in sea water and 30% sea water, respectively. Then the sodium potentials in sea water and 30% sea water are

$$E_{Na} = \frac{RT}{F} \ln \frac{[Na]_o}{[Na]_i}$$

and

$$E'_{Na} = \frac{RT}{F} \ln \frac{[Na]'_o}{[Na]_i}$$

and we have

$$E'_{Na} - E_{Na} = \frac{RT}{F} \ln \frac{[Na]'_o}{[Na]_o}. \tag{2.2}$$

But

$$E_{Na} = V_{Na} + E_r, \tag{2.3}$$

where E_r is the resting potential and V_{Na} is the value of the voltage clamp at which there is no initial ionic current flow in the case of the axon immersed in sea water. Similarly

$$E'_{Na} = V'_{Na} + E'_r, \tag{2.4}$$

where E'_r is the resting potential and V'_{Na} is the value of the voltage clamp at which there is no initial ionic current flow in the case of the axon immersed in 30% sea water. From (2.3) and (2.4), it follows that

$$E'_{Na} - E_{Na} = (V'_{Na} - V_{Na}) + (E'_r - E_r). \tag{2.5}$$

Now (2.2) and (2.5) give independent estimates for $E'_{Na} - E_{Na}$. The expression in (2.2) is a theoretical prediction based on the use of the Nernst formula; the expression in (2.5) is obtained from experimental work. That is, the values V'_{Na}, V_{Na} and E'_r, E_r can all be determined experimentally. The expression in (2.5) is obtained by assuming that (2.3) and (2.4) hold, but they hold if we assume that the initial ionic current is composed of Na ions. Thus, if it turns out that the right-hand sides of (2.2) and (2.5) have the same value, we will have obtained substantial quantitative evidence that the initial ionic current is composed of Na ions. In fact the values obtained for the right-hand sides of (2.2) and (2.5) agree very well. For a comparison of the two values in a number of experiments, see Hodgkin and Huxley (1952a, p. 454).

This leaves us with the question of the composition of the later ionic current. If the voltage is clamped at a value $V > 12$ mV, then the later ionic current is actually in the opposite direction from the initial ionic current (see Fig. 2.4). This suggests that the later ionic current flow consists of ions different from sodium. It turns out that there are cogent qualitative arguments to support the hypothesis that this current consists of K ions. See the discussion in Hodgkin and Huxley (1952a, pp. 455–457).

Since we have discussed the Na ion current in some detail and since our purpose is to convey a general picture rather than a complete set of arguments, we will simply accept the conclusion that the later ionic current is composed mainly of K ions and move to the next step, that is, to obtain a quantitative description of the flow of Na ions and quantitative description of the flow of K ions. We assume that the initial ionic current consists of Na ions and that the later ionic current consists of K ions, but that leaves open the question of the composition of the ionic current at the in-between period during which, presumably, the Na ion flow is decreasing and the K ion flow is increasing. Answering this question presents serious difficulties because there are no experimental techniques for measuring the Na ion current and the K ion current separately. The experimental results yield simply the magnitude and the direction of flow of the ionic current. In order to obtain a quantitative description of the constituents of the ionic current, that is, to obtain graphs of the Na and K ion currents as functions of time, we will make some assumptions about the nature of the ion currents. First we will state these assumptions and then explain their status and how they can be justified. Let $I_K(t)$ and $I_{Na}(t)$ denote the Na and K ion currents as functions of time, and, as before, let the prime sign denote the 30% sea water case.

Assumption 1. The function $I_K(t)$ is independent of $[Na]_o$, the exterior concentration of Na ions.

Assumption 2. If $I_{Na}(t)$ and $I'_{Na}(t)$ denote the Na ion currents as functions of time and if the exterior concentrations of Na ions has the values $[Na]_o$ and $[Na]'_o$, then there exists a constant k such that for all t,

$$\frac{I'_{Na}(t)}{I_{Na}(t)} = k.$$

[We have stated this assumption in mathematical form because that is the form in which it will be applied a little later. The original form of the assumption, as stated by Hodgkin and Huxley (1952a, p. 457), is more tentative. Depending on the point of view, one may say it is vaguer or less rigid.]

Assumption 3. Let $[t_0, t_0 + T]$ denote the interval during which $I_{Na}(t)$ increases to its maximum value. We assume that if $t \in [t_0, t_0 + T/3]$ or $t_0 \leq t \leq t_0 + T/3$, then

$$\frac{dI_K}{dt}(t) = 0.$$

At this stage these assumptions can be advanced on the following basis. The first two assumptions are very simple and they do not conflict with any experimental results. The third assumption is strongly suggested by the experimental data when the value V of the voltage clamp is such that $V + E_r$ is near the sodium potential, that is, if V has such a value (and hence the initial Na ion current is negligible), the total current remains very close to zero for an interval of time. Thus, we have reasons for trying these assumptions. As it turns out, they lead to a consistent theoretical structure and quantitative description that yield good agreement with other experimental evidence. Thus, we may say that these assumptions are justified by the results that they yield.

Now we are ready to determine the functions $I_{Na}(t)$ and $I_K(t)$. In order to do this, we use two voltage-clamp experiments:

(i) A voltage-clamp experiment in which an axon is immersed in sea water and the voltage is clamped at a value V.

(ii) A voltage-clamp experiment in which the same axon is immersed in 30% sea water and the voltage is clamped at the same value V.

Allowance is made for the surge of capacitance current, the deterioration of the axon as a result of having experiments inflicted upon it, and the difference in resting potential of the axon in the two different fluids in which it is immersed. How these allowances are made is discussed in Hodgkin and Huxley (1952a, p. 458). We obtain two functions, $I_i(t)$ and $I_i'(t)$, the ionic currents, as functions of time, across the membrane when the axon is immersed in sea water and 30% sea water, respectively. We assume that each of these ionic currents is the sum of a sodium current and a potassium current, that is, we assume that

$$I_i(t) = I_{Na}(t) + I_K(t) \tag{2.6}$$

and

$$I_i'(t) = I_{Na}'(t) + I_K'(t), \tag{2.7}$$

where $I_{Na}(t)$, $I_{Na}'(t)$ denote sodium currents and $I_K(t)$, $I_K'(t)$ denote potassium currents. Our objective is to use Assumptions 1–3 in order to determine $I_{Na}(t)$, $I_{Na}'(t)$, $I_K(t)$, and $I_K'(t)$. More precisely, we will solve for these functions in terms of $I_i(t)$ and $I_i'(t)$. By Assumption 1, $I_K(t) = I_K'(t)$, and by Assumption 2,

$$I_{Na}'(t) = kI_{Na}(t),$$

where k is a constant that has not yet been determined. Hence, (2.7) becomes

$$I_i'(t) = kI_{Na}(t) + I_K(t). \tag{2.8}$$

Subtracting (2.8) from (2.6) we obtain

$$I_i(t) - I_i'(t) = (1 - k)I_{Na}(t)$$

or

$$I_{Na}(t) = \frac{I_i(t) - I_i'(t)}{1 - k}. \qquad (2.9)$$

Since $I_i(t)$ and $I_i'(t)$ are known, it remains to determine the constant k. In order to do this, the slopes dI_i/dt and dI_i'/dt are measured at $t = 0$, that is, at the beginning of the voltage clamp (remember that the surge of capacitance current has already been allowed for). Let c be the constant defined by $c = dI_i'/dt(0) \, dI_i/dt(0)$. Differentiating (2.6) and (2.7), we obtain

$$\frac{dI_i}{dt} = \frac{dI_{Na}}{dt} + \frac{dI_K}{dt}$$

and

$$\frac{dI_i'}{dt} = \frac{dI_{Na}'}{dt} + \frac{dI_K'}{dt}.$$

By Assumption 1,

$$\frac{dI_K'}{dt} = \frac{dI_K}{dt},$$

and by Assumption 3,

$$\frac{dI_K}{dt}(0) = 0.$$

Hence, we obtain

$$\frac{dI_i}{dt}(0) = \frac{dI_{Na}}{dt}(0), \qquad (2.10)$$

$$\frac{dI_i'}{dt}(0) = \frac{dI_{Na}'}{dt}(0). \qquad (2.11)$$

Dividing (2.11) by (2.10) and using the definition of c, we obtain

$$c = \frac{dI_i'/dt(0)}{dI_i/dt(0)} = \frac{dI_{Na}'/dt(0)}{dI_{Na}/dt(0)}.$$

But by Assumption 2, the ratio $[I_{Na}'(t)]/[I_{Na}(t)]$ is a constant k

and hence that constant is

$$k = c = \frac{dI_i'(0)}{dt} \Big/ \frac{dI_i}{dt}(0).$$

Substituting this value in (2.9), we obtain an explicit solution for $I_{Na}(t)$, and since

$$I_{Na}'(t) = kI_{Na}(t),$$

we obtain the following expression for $I_{Na}'(t)$:

$$I_{Na}'(t) = \frac{k}{1-k}[I_i(t) - I_i'(t)].$$

Finally by (2.6) and the condition $I_K(t) = I_K'(t)$ (i.e., Assumption 1), we have

$$I_K'(t) = I_K(t) = I_i(t) - I_{Na}(t)$$

$$= I_i(t) - \frac{1}{1-k}(I_i(t) - I_i'(t))$$

$$= \frac{(1-k)I_i - I_i + I_i'}{1-k}$$

$$= \frac{I_i' - kI_i}{1-k}.$$

Thus, with the use of Assumptions 1–3, the functions $I_{Na}(t)$, $I_K(t)$, $I_{Na}'(t)$, and $I_K'(t)$ for a fixed voltage clamp can be determined. Examples of these functions are shown in Fig. 2.7.

Knowledge of the sodium and potassium currents [i.e., $I_{Na}(t)$ and $I_K(t)$] is sufficient to explain qualitatively how an action potential propagates; that is, the experimental results we have just described are sufficient to suggest the explanation given in Section 2.2 for how an action potential propagates.

As we have seen from the voltage-clamp experiments, if the membrane is depolarized above the threshold value (i.e., if $V > 12$–15 mV) then Na current begins to flow. Since $[Na]_o > [Na]_i$, this Na current is directed inward and the membrane is depolarized further until the membrane potential becomes positive and approaches the sodium equilibrium potential. Then, as the experi-

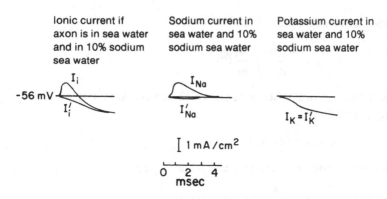

Figure 2.7. Separation of ionic current into I_{Na} and I_K.

ments show, the permeability of the membrane to Na current begins to decrease and the permeability of the membrane to K current begins to increase. Since $[K]_i > [K]_o$, the K current is an outward flow of K ions. This K current brings the membrane potential back to the resting value at which level the ionic current becomes insignificant. Meanwhile the action potential is propagated along the nerve fiber because points on the fiber near the point under observation have their membrane potentials depolarized to the threshold value.

Our next objective is to obtain a quantitative version of the description just given. In order to do this, a number of preliminaries must be carried out.

First, we want to establish an independence principle. This principle states that the probability that any individual ion will cross the membrane in a specified interval of time is independent of any other (chemically different) ions that are present. We will establish the principle by using it to derive an equation that predicts the effect of sodium concentration on sodium current. It can be shown that the predictions of this equation agree with experimental results. This establishes or certainly supports the validity of the independence principle.

If we assume that the probability that any individual ion will cross the membrane in a specified interval of time is independent of other ions that are present, then the inward flux M_1 of a particular ion species will be proportional to the concentration c_1

of that ion in the external bath. Hence, we may write

$$M_1 = k_1 c_1, \tag{2.12}$$

where k_1 is a constant that depends on the condition of the membrane and on the potential difference E across the membrane. We will take E as fixed. Similarly if M_2 is the outward flux,

$$M_2 = k_2 c_2, \tag{2.13}$$

where c_2 is the ion concentration in the interior of the axon and k_2 is a constant. Dividing (2.12) by (2.13), we obtain

$$\frac{M_1}{M_2} = \frac{k_1 c_1}{k_2 c_2}. \tag{2.14}$$

The condition for equilibrium is $M_1 = M_2$. Substituting in (2.14), we obtain a condition for equilibrium:

$$\frac{k_2}{k_1} = \frac{c_1^*}{c_2}, \tag{2.14'}$$

where c_2 is a fixed internal concentration and c_1^* is the external concentration that would be in equilibrium with c_2 under the value E of the membrane potential. Thus by the Nernst formula,

$$E = \frac{RT}{F}\left[-\ln\frac{c_1^*}{c_2} \right]$$

or

$$\frac{c_1^*}{c_2} = \exp\left[-\frac{EF}{RT} \right]. \tag{2.15}$$

Now if E^* is the equilibrium potential for the ion and if c_1 is the external concentration, then (again by the Nernst formula)

$$E^* = \frac{RT}{F}\ln\frac{c_2}{c_1} = -\frac{RT}{F}\ln\frac{c_1}{c_2}$$

or

$$\frac{c_1}{c_2} = \exp\left[-\frac{E^*F}{RT} \right]. \tag{2.16}$$

Dividing (2.16) by (2.15), we obtain

$$\frac{c_1}{c_1^*} = \exp\left[-\frac{(E^* - E)F}{RT}\right].$$

Substituting from (2.14') into (2.14) and applying (2.16), we have

$$\frac{M_1}{M_2} = \frac{c_1}{c_1^*} = \exp\left[-\frac{(E^* - E)F}{RT}\right]. \qquad (2.17)$$

Now we compare the Na currents if the axon is immersed first in sea water with sodium concentration $[Na]_o$ and then in a low-sodium solution with sodium concentration $[Na]_o'$. We assume that the membrane potential has the value E in both cases. Let $M_{Na,1}$ and $M_{Na,2}$ be the inward and outward fluxes, respectively, of sodium when the axon is immersed in sea water. Fluxes $M_{Na,1}'$ and $M_{Na,2}'$ are similarly defined if the axon is immersed in a solution with sodium concentration $[Na]_o'$. Then

$$\frac{I_{Na}'}{I_{Na}} = \frac{M_{Na,1}' - M_{Na,2}'}{M_{Na,1} - M_{Na,2}}. \qquad (2.18)$$

Since

$$M_{Na,1}' = k_1 [Na]_o'$$

and

$$M_{Na,1} = k_1 [Na]_o,$$

then

$$\frac{M_{Na,1}'}{M_{Na,1}} = \frac{[Na]_o'}{[Na]_o}; \qquad (2.19)$$

but

$$M_{Na,2} = k_2 [Na]_i$$

and

$$M_{Na,2}' = k_2 [Na]_i,$$

and thus

$$M_{Na,2} = M_{Na,2}'. \qquad (2.20)$$

Substituting from (2.19), (2.20), and (2.17) into (2.18), we obtain

$$
\frac{I'_{Na}}{I_{Na}} = \frac{M_{Na,1}[Na]'_o/[Na]_o - M_{Na,2}}{M_{Na,1} - M_{Na,2}}
$$

$$
= \frac{\{[Na]'_o/[Na]_o(M_{Na,1}/M_{Na,2}) - 1\}M_{Na,2}}{\{(M_{Na,1}/M_{Na,2}) - 1\}M_{Na,2}}
$$

$$
= \frac{\{[Na]'_o/[Na]_o \exp([-(E^* - E)F]/RT)\} - 1}{\exp([-(E^* - E)F]/RT) - 1}
$$

$$
= \frac{\{[Na]'_o/[Na] \exp([-(E_{Na} - E)F]/RT)\} - 1}{\exp([-(E_{Na} - E)F]/RT) - 1}.
$$

$$(2.21)$$

Equation (2.21) shows how the sodium concentration affects the sodium current. It turns out that the theoretical prediction made by (2.21) agrees very well with experimental results. For a detailed discussion of this, see Hodgkin and Huxley (1952a, pp. 468–471). Thus the independence principle has been established.

Next it turns out that instead of dealing with the ionic currents I_{Na} and I_K themselves, it is considerably more advantageous to consider the permeability of the membrane to Na and K ions. In order to clarify this point we first give a formal definition of permeability.

Definition. The *permeability of the membrane to Na ions is*

$$
\frac{I_{Na}}{E - E_{Na}}.
$$

As previously defined,

$$
E_{Na} = \frac{RT}{F} \ln \frac{[Na]_o}{[Na]_i} = -\frac{RT}{F} \ln \frac{[Na]_i}{[Na]_o}.
$$

E_{Na} is called the sodium equilibrium potential; $-E_{Na}$ is the emf equivalent to the chemical driving force acting on the Na ions. Thus, the effective external emf acting on the Na ions is $E - E_{Na}$.

The permeability is denoted by g_{Na}. It has the dimension of conductance (current divided by potential difference) and g_{Na} will frequently be referred to as the sodium conductance. Similarly, the *permeability of the membrane to K ions* is $I_K/(E - E_K)$ and is denoted by g_K.

Since *permeability* has the dimension of conductance, which is the reciprocal of resistance, the units of measure of permeability are the reciprocals of units of resistance. The basic unit is the mho, which is the reciprocal of the ohm. That is, $1 \; \Omega^{-1}$ is the conductance that exists if 1 A of current flows between two points whose potential difference is 1 V.

The quantities E_{Na}, E, and I_{Na} have already been described and we know how to determine them from experimental data. Hence the function $g_{Na}(t)$ can be determined and graphed. The value of E_K can also be determined experimentally [see Hodgkin and Huxley (1952b)] and I_K has already been described. Hence the function $g_K(t)$ can also be determined and graphed.

There are two important reasons for introducing the concept of permeability. First, it can be shown that g_{Na}, g_K are independent of the magnitudes of the corresponding driving forces $E - E_{Na}$, $E - E_K$ under which g_{Na}, g_K are measured. [This is shown by Hodgkin and Huxley (1952a).] Thus g_{Na}, g_K measure real properties of the membrane just as resistance measures a real property of, say, a length of copper wire. Second, at the sodium potential the current curves reverse their signs, whereas the conductance curves undergo no marked change. Also, the function $g_{Na}(t)$ is better behaved, mathematically. Indeed, $g_{Na}(t)$ is continuous, whereas $I_{Na}(t)$ is discontinuous. [See Hodgkin and Huxley (1952a, p. 477).] Consequently, g_{Na} and g_K play primary roles in the Hodgkin–Huxley equations that will presently be derived.

In order to obtain the desired quantitative description of how the action potential is propagated, considerably more quantitative data from voltage-clamp experiments is needed. The voltage-clamp experiments described so far concern conditions in which the membrane potential is clamped or fixed at a value different from the resting value. That is, we obtained a description of the currents $I_{Na}(t)$ and $I_K(t)$ across the membrane when $V \neq 0$. To obtain the quantitative description of how the action potential is propagated, we need descriptions of the currents $I_{Na}(t)$ and $I_K(t)$ and the corresponding permeabilities $g_{Na}(t)$ and $g_K(t)$ in other voltage-clamp conditions: First the case in the membrane potential is clamped at a value different from the resting value (i.e., $V \neq 0$) and then suddenly restored to the resting value ($V = 0$). Also necessary are certain voltage-clamp experiments in which the membrane potential is clamped at a value different from the resting value (i.e., $V = V_1 \neq 0$) and then suddenly changed to a new value different from the resting value (i.e., $V = V_2$ and $V_2 \neq 0$ and $V_2 \neq V_1$). Experiments of this kind are described in Hodgkin and Huxley (1952b). Finally a more detailed quantitative experimental study is needed on how the sodium conductance decreases after reaching its initial maximum. Such experiments are described in Hodgkin and Huxley (1952c). Since our purpose is to sketch the experimental procedures and results rather than to give a detailed account, we will not describe the experiments in Hodgkin and Huxley (1952b, 1952c). However, we will feel free to invoke the results of these experiments when we derive the Hodgkin–Huxley equations. For the present we will just enumerate a few of the results obtained in Hodgkin and Huxley (1952b).

First the potassium equilibrium potential E_K is determined. Second, the permeabilities $g_{Na}(t)$, $g_K(t)$ are shown to be independent of the driving force. Finally, besides the two main ionic currents $I_{Na}(t)$ and $I_K(t)$, there is an appreciable "leakage" current due mainly to a flow of chloride ions, which is discussed and estimated quantitatively in Hodgkin and Huxley (1952b).

2.3.2 Derivation of the differential equations

If the experimental data are compared with the previous description of the propagation of the action potential, it becomes clear that in obtaining this description we have not made full use of the experimental data. That is, the experimental data consist of specific numbers, whereas the description of the action potential is qualitative, not quantitative. The description contains such phrases as "begins to flow," "depolarize further," and "begins to decrease." But these are not numbers in the description. Clearly then we should look for a quantitative description, that is, a description that uses the full force of the quantitative experimental results. Moreover, our quantitative description should not just specify quantities but should be expected also to yield quantitative explanations: for example, we would expect to obtain a mathematical relationship among the membrane potential, the sodium conductance, and the potassium conductance. Also we need a considerably more detailed description of the action potential than the one already given. (For example, so far we have no explanation of how the absolute and relative refractory periods occur.) This is the kind of description that Hodgkin and Huxley (1952c) obtained in their paper. They used the experimental results to derive a system of differential equations in which the dependent variables or unknowns are the potential difference across the membrane as a function of time and a set of variables that describe the sodium and potassium conductances as functions of time.

Our next step is to derive these Hodgkin–Huxley differential equations. Before starting the actual derivation, however, it is important to clarify the status of these equations. This is especially important to the reader who is accustomed to differential equations in physics, for example, mechanics or electrical circuit theory. In deriving a differential equation or system of differential equations in physics, one usually starts from first principles such as Newton's laws in mechanics or Kirchhoff's laws in electrical circuit theory. The concepts in the general principles are expressed in appropriate quantitative terms. For example, in linear motion, the acceleration is represented by d^2x/dt^2. Finally, the particular given conditions are expressed in mathematical terms. A familiar example is the derivation of the equation of a simple pendulum. One starts from Newton's Second Law: Force equals mass times acceleration. If m denotes the mass of the bob, l the length of the pendulum, and θ the angle of displacement, then

$$\text{mass} \times \text{acceleration} = ml\frac{d^2\theta}{dt^2}.$$

The force that acts on the bob is a component of the gravitational

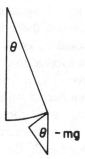

Figure 2.8.

force $-mg$ and, as indicated in Fig. 2.8, that force is $-mg \sin \theta$. Thus the differential equation for the simple pendulum is

$$ml\frac{d^2\theta}{dt^2} + mg \sin \theta = 0.$$

The derivation of the Hodgkin–Huxley equations is in stark contrast to this familiar procedure. The chief reason for this is that there are no basic laws or first principles that we can follow, that is, there are no Newton's laws or Kirchhoff's laws from which to start. Experimental work suggests that the membrane potential and the sodium current and the potassium current are the most important variables and we would expect, therefore, that the equations might take the form:

$$\frac{dV}{dt} = F(t, V, g_{Na}, g_K),$$

$$\frac{dg_{Na}}{dt} = G(t, V, g_{Na}, g_K),$$

$$\frac{dg_K}{dt} = H(t, V, g_{Na}, g_K),$$

and the derivation of the equations would consist in determining the functions F, G, and H. [As we will see later, we do not deal directly with g_{Na} and g_K. Instead special forms are chosen for the functions $g_{Na}(t)$ and $g_K(t)$.] But we have no theoretical guidance or first principles to help us determine the functions F, G, and H. Instead, the form of the differential equations is determined entirely from the experimental data. The differential equations are empirical or ad hoc descriptions. Having indicated the nature of the equations, we list several points that should be emphasized. First and most important is the fact that we are forced to consider ad hoc equations because there are no first principles to guide us. The quantitative aspects of physiology have not been developed to the stage where such first principles exist; and, indeed, there is no particular reason to believe that a set of first principles like Newton's laws or Kirchhoff's laws will be developed. The future development of quantitative biology remains veiled. All we know at present is that there is a significant quantitative aspect of biology. Second, the Hodgkin–Huxley equations turn out to be (as we will show later) a very good summary or fusion of quantitative experimental results. Their empirical quality makes them analogous to Kepler's laws. To some extent this analogy suggests the status of mathematical biology at the present. Since extensive experimental data can be represented in compact mathematical form, that is, a system of differential equations, it seems clear that further mathematical developments can be expected. On the other hand, the analogy of the Hodgkin–Huxley equations with Kepler's laws cannot be pushed too far. There is no reason to conclude from the analogy that a theoretical biologist will presently develop some basic principles that will explain the origin of the Hodgkin–Huxley equations just as Newton's laws can be used to derive or explain Kepler's laws. (It is possibly worth noting that the fact that the mathematical description given by Hodgkin and Huxley is in the form of differential equations is misleading to a reader who is familiar with mathematical physics. With a background of physics, one tends automatically to regard a differential equation as a representation or realization of basic physical laws. The Hodgkin–Huxley equations have no such properties, but correspond instead to Kepler's laws, which, of course, are not differential equations.)

Third, it is enlightening to emphasize that Hodgkin and Huxley were fully aware of the ad hoc nature of their differential equations. As physiologists, their objective was to determine how the action potential occurs and is propagated. (As we have pointed out before, even today the basic question of how the flows of Na and K ions are regulated remains unanswered.) When Hodgkin and Huxley initiated their work, they were not looking for some differential equations. They were concerned with studying the flow of electric current through the membrane surface and finding an explanation for how these currents occur. As it turned out, the hypothesis that they started out to test in their experiments proved to be incorrect; even the hope of formulating another hypothesis that would explain the nature of the flow of current failed. It had seemed reasonable that with the large amount of experimental data, it would be possible to develop such a molecular hypothesis, but Hodgkin and Huxley were forced to conclude that their experimental data would "yield only very general information about the class of systems likely to be involved" [Hodgkin (1977, p. 19)]. Thus they "settled for the more pedestrian aim of finding a simple set of mathematical equations which might plausibly represent the movement of electrically charged gating particles" [Hodgkin (1977, p. 19)].

Finally, it is important to understand the status of the Hodgkin–Huxley equations in the view of physiologists. This status is clearly described in the following excerpt from Hodgkin and Huxley (1952d, p. 541):

> The agreement must not be taken as evidence that our equations are anything more than an empirical description of the time-course of the changes in permeability to sodium and potassium. An equally satisfactory description of the voltage clamp data could no doubt have been achieved with equations of very different form, which would probably have been equally successful in predicting the electrical behavior of the membrane. It was pointed out in Part II of this paper that certain features of our equations were capable of a physical interpretation, but the success of the equations is no evidence in favour of the mechanism of permeability change that we tentatively had in mind when formulating them.
>
> The point that we do consider to be established is that fairly simple permeability changes in response to alterations in membrane potential, of the kind deduced from the voltage clamp results, are a sufficient explanation of the wide range of phenomena that have been fitted by solutions of the equations.

The Hodgkin–Huxley equations come in part from the statement concerning current that has already been invoked several times, that is, the equation

$$I = C_M \frac{dV}{dt} + I_i, \tag{2.22}$$

which says that the total current is the sum of the capacitance current and the ionic currents. The capacitance of the membrane per unit area (i.e., C_M) has already been determined. There are three ionic currents: The two most important are I_{Na} and I_K, the sodium and potassium currents that we have described in some detail, and a third "leakage" current, denoted by I_l, consisting largely of chloride ions, which was already mentioned briefly. The leakage current is discussed in Hodgkin and Huxley (1952c). Thus

$$I_i = I_{Na} + I_K + I_l \tag{2.23}$$

and, substituting (2.23) into (2.22), we have

$$I = C_M \frac{dV}{dt} + I_{Na} + I_K + I_l. \tag{2.24}$$

Using the independence principle discussed earlier, we regard I_{Na}, I_K, and I_l as having no influence on one another. Hence we may deal with the sodium and potassium currents separately and write

$$I_{Na} = g_{Na}(E - E_{Na}),$$
$$I_K = g_K(E - E_K).$$

It turns out that the leakage current has an especially simple description

$$I_l = \bar{g}_l(E - E_l),$$

where $\bar{g}_l$ is a positive constant and E_l is the equilibrium potential for the ions (mostly chloride) that constitute the leakage current. Thus (2.24) becomes

$$I = C_M \frac{dV}{dt} + g_{Na}(E - E_{Na}) + g_K(E - E_K) + \bar{g}_l(E - E_l) \tag{2.25}$$

or

$$\frac{dV}{dt} = \frac{1}{C_M}[I - g_{Na}(E - E_{Na}) - g_K(E - E_K) - \bar{g}_l(E - E_l)]. \tag{2.26}$$

Since

$$V = E - E_R,$$

it is convenient to rewrite $E - E_{Na}$, $E - E_K$, and $E - E_l$ in terms of V. To do thus, let

$$V_{Na} = E_{Na} - E_r,$$
$$V_K = E_K - E_r,$$
$$V_l = E_l - E_r.$$

Then

$$E - E_{Na} = V - V_{Na},$$
$$E - E_K = V - V_K,$$
$$E - E_l = V - V_l,$$

and (2.26) becomes

$$\frac{dV}{dt} = \frac{1}{C_M} \left[I - g_{Na}(V - V_{Na}) - g_K(V - V_K) - \bar{g}_l(V - V_l) \right].$$

$$(2.27)$$

Equation (2.27) is a version of the first of the Hodgkin–Huxley equations. It would be reasonable that the remaining equations would be of the form

$$\frac{dg_{Na}}{dt} = G(t, V, g_{Na}, g_K),$$

$$\frac{dg_K}{dt} = H(t, V, g_{Na}, g_K),$$

and our remaining problem would be to determine the functions G and H from the experimental data. In fact, the problem is by no means that straightforward. Hodgkin and Huxley decided instead to introduce other variables. The choice of these variables cannot be entirely explained on a logical basis. All we can do is point out reasons that suggest these choices for the variables. The decision to use these variables was a brilliant intuitive step. A measure of the extraordinary accomplishment that this step represents is the fact that the mathematical formulation given by this choice of variables has remained essentially unchanged in the 30 years of work on neurons and other electrically excitable cells that have followed.

We look first at the simpler case, that is, the potassium conductance g_K, and we examine how $g_K(t)$ changes when V is changed

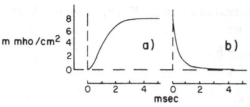

(a) Rise of potassium conductance associated with depolarization of 25 mV

(b) Fall of potassium conductance associated with repolarization to the resting potential

Figure 2.9. Graphs of potassium conductance.

from 0 to 25 mV and then from 25 mV back to 0. This is displayed in Fig. 2.9. One of the complications that is shown by Fig. 2.9 is the fact that if V is changed from 0 to 25 mV, then $g_K(t)$ rises with a marked inflection. But if V is changed from 25 mV to 0, then $g_K(t)$ decreases in a simple exponential way. Consequently, finding a differential equation whose solutions would fit the two parts of the curve in Fig. 2.9 is complicated. In order to obtain a valid description of this kind of curve, Hodgkin and Huxley introduced a new dimensionless variable n. They proposed to express the function $g_K(t)$ as

$$g_K(t) = \bar{g}_K[n(t)]^4, \tag{2.28}$$

where $\bar{g}_K$ is a positive constant whose value will be obtained from the experimental data, and to require that the function $n(t)$ satisfy a differential equation of the form

$$\frac{dn}{dt} = \alpha_n(1-n) - \beta_n n, \tag{2.29}$$

where α_n and β_n are nonnegative functions of V. The best justification for this proposal and a similar, somewhat more complicated, proposal made for $g_{Na}(t)$ is the fact that it works. An argument that suggests introducing the description of $g_K(t)$ given by (2.28) and (2.29) is the following. If α_n and β_n are constants, then by an elementary method (variables separable) (2.29) can be solved and we obtain, for the general solution,

$$n(t) = \left[\frac{\alpha_n}{\alpha_n + \beta_n}\right] - K\exp[-(\alpha_n + \beta_n)t], \tag{2.30}$$

where K is a positive constant. If for simplicity we take $K = 1$, then if the solution $n(t)$ is close to zero, it can be approximated by taking $\beta_n = 0$ in (2.30) and we obtain

$$n(t) = -e^{-t} + 1.$$

A simple calculation shows that the function

$$[n(t)]^4 = [1 - e^{-t}]^4$$

has an inflection point just as appears in Fig. 2.9(a). Similarly if $|n(t)|$ has a value close to 1, (2.30) can be approximated by taking $\alpha_n = 0$ and $K = 1$, $\beta_n = 1$. We obtain

$$n(t) = -e^{-t},$$

and the function

$$[n(t)]^4 = [-e^{-t}]^4 = e^{-4t}$$

describes a simple exponential decrease just as appears in Fig. 2.9(b).

Assuming that (2.28) and (2.29) give a satisfactory description of $g_K(t)$, we must proceed to determine α_n and β_n as functions of V. In order to do this, we must, of course, utilize the experimental results.

Let n_0 denote the resting value of $n(t)$, that is, n_0 is the number such that

$$\bar{g}_K(n_0)^4$$

is the potassium conductance if $V = 0$ (i.e., the potassium conductance of the membrane if the membrane potential is equal to the resting potential). Note that since we have not specified $\bar{g}_K$ ($\bar{g}_K$ will be determined later from the experimental data), we do not have an explicit value for n_0. But since $\bar{g}_K$ can be determined independently from the experimental data, n_0 has a definite value that we can obtain from the experimental data. Now suppose a voltage-clamp experiment is performed in which V is changed suddenly from 0 to a nonzero value. Then α_n, β_n are changed to values that correspond to the new nonzero value of V. With α_n, β_n at these new fixed values, (2.29) can be solved and we obtain

$$\ln\left[\frac{\alpha_n}{\alpha_n + \beta_n} - n\right] = -(\alpha_n + \beta_n)t + C, \tag{2.31}$$

where C is a constant of integration. Let

$$n_\infty = \frac{\alpha_n}{\alpha_n + \beta_n}. \tag{2.32}$$

Note that if $n = n_\infty$, then by (2.29), $dn/dt = 0$. Hence n_∞ is such that $\bar{g}_K n_\infty^4$ is the value of the potassium conductance ultimately obtained after V is clamped at the nonzero value. (Remember that the experimental data show that following imposition of a voltage clamp, the potassium conductance increases to a certain level where it then remains fixed.) Let us denote this final value of potassium conductance by g_{K_∞} and observed that it is a value that can be obtained from this experimental data. Substituting from (2.32) into (2.30) and simplifying, we obtain

$$n(t) = n_\infty - K \exp[-(\alpha_n + \beta_n)t].$$

If we impose the initial condition

$$n(0) = n_0,$$

we obtain

$$n(t) = n_\infty - (n_\infty - n_0)\exp[-(\alpha_n + \beta_n)t].$$

Introducing the notation $\tau_n = 1/(\alpha_n + \beta_n)$, we obtain for the solution

$$n(t) = n_\infty - (n_\infty - n_0)\exp\left[-\frac{t}{\tau_n}\right].$$

Thus

$$[g_K(t)]^{1/4} = (\bar{g}_K)^{1/4} n(t)$$

$$= (\bar{g}_K)^{1/4} n_\infty - (\bar{g}_K)^{1/4}(n_\infty - n_0)\exp\left[-\frac{t}{\tau_n}\right].$$

Since $\bar{g}_K n_\infty^4 = g_{K_\infty}$, then

$$[g_K(t)]^{1/4} = (g_{K_\infty})^{1/4} - \left[(g_{K_\infty})^{1/4} - (\bar{g}_K)^{1/4}(n_0)\right]\exp\left[-\frac{t}{\tau_n}\right]. \tag{2.33}$$

But $\bar{g}_K n_0^4$ is by definition the potassium conductance when $V = 0$ and consequently the value of $\bar{g}_K n_0^4$ is known from the experimental data; so if we denote $\bar{g}_K n_0^4$ by g_{K_0}, then we may rewrite (2.33) as

$$g_K(t) = \left\{ (g_{K_\infty})^{1/4} - \left[(g_{K_\infty})^{1/4} - (g_{K_0})^{1/4} \right] \exp\left[-\frac{t}{\tau_n} \right] \right\}^4.$$

(2.34)

Since g_{K_∞} and g_{K_0} are known from the experimental data, then if a fixed value is chosen for τ_n, the right-hand side of (2.34) is a specific function of t. On the other hand, for fixed V, the function $g_K(t)$ has been determined experimentally. Consequently in order to obtain the correct value for τ_n, we simply select the value of τ_n so that the graph of the resulting $g_K(t)$, which is defined by (2.34), agrees with or coincides with the experimental graph for $g_K(t)$. Having thus determined τ_n, we have an equation relating α_n and β_n, that is,

$$\frac{1}{\alpha_n + \beta_n} = \tau_n$$

or

$$\alpha_n \tau_n + \beta_n \tau_n = 1.$$

(2.35)

Next we consider the problem of estimating $\bar{g}_K$. Experimental results show that if V has a large value (i.e., a value > 110 mV), then the corresponding value of g_{K_∞} takes values ~ 20–50% higher than the value g_{K_∞} corresponding to $V = 100$ mV. For definiteness in calculation, we assume that the values are 20% higher, and we assume that $n = 1$ at those values. Since $g_{K_\infty} = 20$ mΩ^{-1}/cm^2 at $V = 100$ mV, then $\bar{g}_K$ is chosen to be 24 mΩ^{-1}/cm^2. (There is, of course, an arbitrariness in this choice of $\bar{g}_K$. This arbitrariness is justified in the same way that some of the earlier assumptions are justified: They contribute to the development of a quantitative description that yields good agreement with other experimental results.)

Now as remarked after (2.32), if V is fixed and sufficient time has elapsed afterward, we have

$$g_K = \bar{g}_K (n_\infty)^4.$$

But $\bar{g}_K$ has been determined and, if V is fixed, then g_K can be determined experimentally as was shown earlier. Hence for a fixed value of V, the quantity n_∞ can be determined. From its definition we have

$$\frac{\alpha_n}{\alpha_n + \beta_n} = n_\infty. \tag{2.36}$$

Now solving (2.35) and (2.36) simultaneously for α_n and β_n, we obtain

$$\alpha_n = \frac{n_\infty}{\tau_n},$$

$$\beta_n = \frac{1 - n_\infty}{\tau_n}.$$

Thus we have computed the values of α_n and β_n associated with a fixed value of V. Values of α_n and β_n that are associated with various values of V are tabulated in Hodgkin and Huxley (1952d, p. 509). It remains to choose functions $\alpha_n(V)$ and $\beta_n(V)$, the graphs of which fit the tabulated values. The functions that Hodgkin and Huxley chose are

$$\alpha_n(V) = \frac{0.01(V+10)}{\exp[(V+10)/10] - 1},$$

$$\beta_n(V) = 0.125 \exp(V/80).$$

Even when variations with temperature and variation of resting potential with $[\text{Na}]_0$ are taken into account, there is still some arbitrariness left in the selection of the functions $\alpha_n(V)$ and $\beta_n(V)$. For a discussion of these points, see Hodgkin and Huxley (1952d, p. 510).

As an indication of the arbitrariness in the equation for n that has been introduced, we point out a remark made by Hodgkin and Huxley in connection with determining the value of τ_n when V is given.

> It will be seen that there is reasonable agreement between theoretical and experimental curves [the graph of the function $g_K(t)$ defined by (2.34) and the experimentally obtained graph of potassium conductance (Fig. 2.9)] except that the latter show more initial delay. Better agreement might have been obtained with a fifth or sixth power [of n] but the *improvement was not considered to be worth the additional complication.* [Italics added.]

The "additional complications" to which Hodgkin and Huxley refer consist mainly in the complications encountered in solving the resulting set of differential equations. It must be remembered that

Hodgkin and Huxley were writing at a time when numerical analysis of a system of differential equations was carried out by lengthy labors with a hand-operated calculating machine [see Hodgkin (1977, p. 19)]. One may well ask whether Hodgkin and Huxley would have proposed a different equation for $n(t)$ if they had had access to a modern computer.

The study of sodium conductance, that is, the function $g_{Na}(t)$, is analogous to the study of potassium conductance except that it is somewhat more complicated. In particular if V is fixed at a value above the threshold value, then the sodium conductance first rises to a maximum value and then decreases. By considerations similar to those used to arrive at the description of potassium conductance given by (2.28) and (2.29), Hodgkin and Huxley decided to describe the sodium conductance by means of the following equations:

$$g_{Na} = m^3 h \bar{g}_{Na}, \tag{2.37}$$

$$\frac{dm}{dt} = \alpha_m(1 - m) - \beta_m m, \tag{2.38}$$

$$\frac{dh}{dt} = \alpha_h(1 - h) - \beta_h h, \tag{2.39}$$

where $\bar{g}_{Na}$ is a positive constant and $\alpha_m, \beta_m, \alpha_h, \beta_h$ are certain nonnegative functions of V. The dependent variables m and h are called the activation and inactivation variables, respectively. They can be regarded as measures of the way by which the membrane permits Na ions to pass through.

By techniques analogous to those used to study α_n and β_n, the coefficients in (2.29), the coefficients $\alpha_m, \beta_m, \alpha_h, \beta_h$ are determined to be

$$\alpha_m = \frac{(0.1)(V + 25)}{\exp[(V + 25)/10] - 1},$$

$$\beta_m = 4 \exp\left[\frac{V}{18}\right],$$

$$\alpha_h = 0.07 \exp\left(\frac{V}{20}\right),$$

$$\beta_h = \frac{1}{\exp[(V + 30)/10] + 1}.$$

Substituting from (2.28) and (2.37) into (2.27), we thus obtain for the Hodgkin–Huxley equations,

$$\frac{dV}{dt} = \frac{1}{C_M}\left[I - \bar{g}_{\text{Na}}m^3h(V - V_{\text{Na}}) - \bar{g}_{\text{K}}n^4(V - V_{\text{K}}) - \bar{g}_l(V - V_l)\right],$$

$$\frac{dm}{dt} = \alpha_m(1 - m) - \beta_m m,$$

$$\frac{dh}{dt} = \alpha_h(1 - h) - \beta_h h,$$

$$\frac{dn}{dt} = \alpha_n(1 - n) - \beta_n n,$$

where $\alpha_m, \beta_m, \alpha_h, \beta_h, \alpha_n, \beta_n$ are the functions of V already described. The preceding equations are the formulation of the Hodgkin–Huxley equations that we will study. For completeness, however, it should be pointed out these equations are valid at the standard temperature of 6.3°C. It has already been mentioned that temperature affects the electrical activity of the axon. A more general mathematical description that includes the effects of temperature T (in degrees Centigrade) is given by

$$\frac{dV}{dt} = \frac{1}{C_M}\left[I - \bar{g}_{\text{Na}}m^3h(V - V_{\text{Na}}) - \bar{g}_{\text{K}}n^4(V - V_{\text{K}}) - \bar{g}_l(V - V_l)\right],$$

$$\frac{dm}{dt} = \left[\alpha_m(1 - m) - \beta_m m\right]\exp\left[(\ln 3)\left(\frac{T - 6.3}{10}\right)\right],$$

$$\frac{dh}{dt} = \left[\alpha_h(1 - h) - \beta_h h\right]\exp\left[(\ln 3)\left(\frac{T - 6.3}{10}\right)\right],$$

$$\frac{dn}{dt} = \left[\alpha_n(1 - n) - \beta_n n\right]\exp\left[(\ln 3)\left(\frac{T - 6.3}{10}\right)\right].$$

That is, the right-hand sides of the equations for the activation and inactivation variables are multiplied by

$$\exp\left[(\ln 3)\left(\frac{T - 6.3}{10}\right)\right] = 3^{[T - (6.3)]/10}.$$

2.4   **What the Hodgkin–Huxley equations describe**

Before going ahead with mathematical analysis of the Hodgkin–Huxley equations, we emphasize what the equations describe. The Hodgkin–Huxley equations can be regarded as describing the variations of membrane potential and ion conductances that occur "naturally" at a fixed point on the axon. (For example, as an action potential passes a given point on the axon, certain changes take place in the membrane potential and the ion conductances.) The equations are obtained from data given by voltage-clamp experiments and are primarily a quantitative description of the results of experiments, that is, the voltage-clamp experiments, which are entirely artificial. That is, in voltage-clamp experiments the membrane potential and the ionic conductances are independent of position on the axon, the membrane potential is constant, and the ionic conductances depend only on time. These are conditions that do not occur in "real life" or the normal functioning of the axon. Later we will see how to extend the description given by the Hodgkin–Huxley equations so that it will include the possibility of variation of membrane potential and ionic conductance with position along the axon. This extended description will thus be broad enough to include description of the propagation of an action potential along the axon.

2.5   **Mathematical status of the Hodgkin–Huxley equations**

2.5.1   **History**

As a mathematical entity, the Hodgkin–Huxley equations have a curious history. They form a significant part of the distinguished physiological work of Hodgkin and Huxley. The importance of this physiological work was recognized immediately; other experimental studies along similar lines were soon carried out by other research scientists, and Hodgkin and Huxley received a Nobel prize for their work only 11 years after the papers appeared. (By the standards of physics and chemistry, this is a fairly long interval, but compared to the intervals for many Nobel prizes in physiology and medicine, 11 years is a very short time.) Hodgkin and Huxley included in their original papers considerable numerical analysis of the equations and in the ensuing years, a number of physiologists have made further numerical analysis of the equations. Even today it seems safe to say that the majority of the physiologically significant mathematical analysis of the equations that has been carried out is numerical analysis.

Nevertheless, from the point of view of physiology a more qualitative analysis of the solutions of the equations is also highly desirable. Throughout our discussion of the derivation of the Hodgkin–Huxley equations we have emphasized the uncertainties in numerical values and the tentativeness of the assumptions that must be made. Consequently it is fairly clear that a study of how solutions of equations of this kind behave in general is just as important as a numerical study of solutions of a particular set of equations. This last observation might be regarded as originating with a mathematician who is looking for work. But in fact the value of qualitative analysis of classes of equations has been emphasized by physiologists [Jack, Noble, and Tsien (1975)] and physicists [Scott (1975)].

In view of these facts, it might be expected that the Hodgkin–Huxley equations would have been the subject of considerable attention by applied mathematicians. This is, however, not the case. Since we are not primarily concerned here with the history of science, we will not speculate too much about why the Hodgkin–Huxley equations were neglected by mathematicians. Among the possible reasons may be the timing; they appeared at a time when nonlinear ordinary differential equations were not at the zenith of their popularity with mathematicians. Moreover at that time (1952), there was a tendency to study very general classes of equations rather than specific systems of equations. (The great interest in recent years in the Lorenz equations shows how values have changed since then.) Perhaps the most important reason is that the Hodgkin–Huxley equations are difficult to study. This difficulty has two aspects. First, although the functions that appear in the equations are sufficiently well behaved (in fact, they are analytic, i.e., can be represented by power series) so that standard existence theorems can be applied to them, the equations are nevertheless a four-dimensional system that is strongly nonlinear. Previous work in nonlinear differential equations, ever since Poincaré, suggests that analysis of these equations would be difficult and that suggestion is entirely correct. There is, however, another aspect of the difficulty with the Hodgkin–Huxley equations that is even more serious, but it is also a difficulty that should make the Hodgkin–Huxley equations of great interest to mathematicians, both pure and applied. The difficulty is that the problems arising with the Hodgkin–Huxley equations that are of physiological interest represent a new direction in the study of nonlinear differential equations. The classical problems in nonlinear ordinary differential equations that arose in celestial mechanics and later in electrical circuit theory are concerned with equilibrium points and then stability, periodic solutions, and almost periodic solutions. [Besides his interest in periodic solutions for problems in celestial mechanics, Poincaré regarded the study of periodic solutions as a crucial first step in the general study of solutions of nonlinear ordinary differential equations. See Poincaré, (1892–99, Vol. I, pp. 81–82).] Efforts to deal with these questions have led to the development of a large amount of mathematical theory.

The problems that arise in nerve conduction theory lie in a different and novel direction. From the point of view of the preceding classical problems, the Hodgkin–Huxley equations are not very interesting. The underlying physiology shows that there should be a unique set of equilibrium values $V = 0$ (which corresponds to the resting potential), m_∞, h_∞, and n_∞. Moreover, if one or more of the variables V, m, h, n is displaced from the equilibrium value, then, again from the physiology, we know that after some time has elapsed, the values of V, m, h, n will return to the equilibrium values. In strict mathematical language we say that if $(\overline{V}, \overline{m}, \overline{h}, \overline{n})$ is a point in R^4 such that

$$(V, m, h, n) \neq (0, m_\infty, h_\infty, n_\infty)$$

and if $(V(t), m(t), h(t), n(t))$ is the solution of the Hodgkin–Huxley equations such that

$$V(0) = \overline{V},$$
$$m(0) = \overline{m},$$
$$h(0) = \overline{h},$$
$$n(0) = \overline{n},$$

then the solution $(V(t), m(t), h(t), n(t))$ is defined for all $t > 0$

and

$$\lim_{t \to \infty} (V(t), m(t), h(t), n(t)) = (0, m_\infty, h_\infty, n_\infty).$$

From the physiological viewpoint, there is a limitation on the initial value $(\overline{V}, \overline{m}, \overline{h}, \overline{n})$. First, as pointed out earlier, we are interested only in values of m, h, n that are in the interval $[0,1]$. Second, we are interested only in values of V that do not damage the axon. If the axon is "fried" by a large value of V, then it would no longer be described by the Hodgkin–Huxley equations. In a little less precise language, all physiologically significant solutions approach the equilibrium point $(0, m_\infty, h_\infty, n_\infty)$. Using conventional mathematical language, we would say that the problem is to prove that there exist two numbers V_1, V_2 such that $V_1 < V_2$ and that the equilibrium point $(0, m_\infty, h_\infty, n_\infty)$ has a region of global asymptotic stability of the form

$$\{(V, m, h, n)/V_1 \le V \le V_2, 0 \le m \le 1, 0 \le h \le 1, 0 \le n \le 1\}.$$

This seems like a reasonable preliminary result to establish, and it is phrased in the language of classical theory of nonlinear ordinary differential equations. Now let us suppose that this result has been established. It has two drawbacks. First, it suggests that from the viewpoint of classical theory, the Hodgkin–Huxley equations are not very interesting. All the solutions approach an equilibrium point. From the classical viewpoint there seems little more to say unless we make a finer investigation and study, for example, the rate at which the solutions approach the equilibrium point. But the result has a more serious drawback: Although it seems to be a complete result, it tells us nothing about whether the solutions describe such phenomena as the refractory period and threshold behavior. We need to use mathematical concepts that correspond to the physiological notions of refractory period and threshold, and we need to establish that the solutions of the Hodgkin–Huxley equations behave in accordance with these mathematical concepts. In other words, we need to develop another mathematical approach. When the problems in nonlinear differential equations that arise in nonlinear electrical circuit theory were first studied intensively in the 1920s, it was realized after a while that much of the theory of ordinary differential equations that had been developed in response to problems in celestial mechanics (for example, the Poincaré–Bendixson theory) could be utilized to study the electrical circuit theory problems. But we have now seen there is no such lucky circumstance when we look at problems in nerve conduction. We cannot simply fall into the familiar language and concepts. From the point of view of the mathematician, the attempt to study quantitatively the mechanism of nerve conduction opens the possibility of a novel study of that most classical and conventional of mathematical objects, the ordinary differential equation.

It is rather surprising that mathematicians have not paid much attention to this possibility. Indeed it was pointed out originally by a physiologist, FitzHugh (1969), rather than a mathematician. Of course, as we shall see, there is considerable literature concerning analysis of the Hodgkin–Huxley equations, and there are effective mathematical tools, such as singular perturbation theory, that can be used to approach this study. But a complete rigorous qualitative analysis has not yet been achieved.

2.5.2 Some successful numerical analysis of the Hodgkin–Huxley equations

2.5.2.1 *Analysis of the original Hodgkin–Huxley equations*

Hodgkin and Huxley, in their original paper (1952d), undertook extensive numerical analysis of the equations they had derived. Their results were largely successful in that the (approximate) numerical solutions of the equations agreed very well with experimental results. Some of these agreements are to be expected. For example, since the equations were derived from data obtained from voltage-clamp experiments, the solutions of the equations might well be expected to give agreements with the results of voltage-clamp experiments. Some of the theoretical agreements or predictions are striking (e.g., agreement with the results of current-clamp experiments, which will be described later) and one of the predictions, the prediction of the velocity of the action potential, is truly spectacular. However, none of the theoretical studies makes a true "prediction" in the sense of giving us new information that had not been previously obtained from experiment or giving a new viewpoint that is suggestive of new directions for experimental work. There are some exceptions to this statement, for example, Guttman, Lewis, and Rinzel (1980).

Because the agreement or predictions given by numerical analysis of the Hodgkin–Huxley equations is such an important measure of the success of the equations, we describe these agreements now in some detail.

First and most prosaic is the correct prediction of the total current during a voltage clamp. During a voltage clamp, V is constant. Consequently $\alpha_m, \beta_m, \alpha_h, \beta_h, \alpha_n, \beta_n$ are constants and, using the notation introduced in Section 2.3, we have

$$m(t) = m_\infty - (m_\infty - m_0)\exp[-(\alpha_m + \beta_m)t],$$
$$h(t) = h_\infty - (h_\infty - h_0)\exp[-(\alpha_h + \beta_h)t],$$
$$n(t) = n_\infty - (n_\infty - n_0)\exp[-(\alpha_n + \beta_n)t],$$

and consequently we can compute the currents

$$I_{Na}(t) = \bar{g}_{Na}m^3h(V - V_{Na}),$$
$$I_K(t) = \bar{g}_K n^4(V - V_K).$$

Since V is a given constant, the leakage current

$$I_l = \bar{g}_l(V - V_l)$$

can also be computed, and thus the total membrane current

$$I = I_{Na} + I_K + I_l$$

can be computed theoretically and compared with the experimentally observed current. It turns out that the theoretical and experimental results agree well unless $V = -115$ mV (the sodium equilibrium potential); then the theoretical current has too short a delay, which is caused by the insufficient delay in the theoretical description of the rise in the potassium conductance (see Fig. 2.9).

Second, a satisfactory agreement is obtained between the experimental results when a shock is applied to an electrode in the axon, and the theoretical results calculated by using the Hodgkin–Huxley equations. The experimental methods were described earlier and the experimental results are summarized in Fig. 2.2. The theoretical result is calculated as follows. Since the potential is constant along the axon, there is no current in the longitudinal direction (i.e., in the direction of the axis of the cylinder); hence the net membrane current I is zero except during the shock. The form of the action potential (we assume that the shock is large enough so that V is raised above the threshold value) can be determined by solving the Hodgkin–Huxley equations with $I = 0$, that is,

$$\frac{dV}{dt} = -\frac{1}{C_M}\left\{ \bar{g}_{Na}m^3h(V - V_{Na}) + \bar{g}_K n^4(V - V_K) + \bar{g}_l(V - V_l) \right\},$$

$$\frac{dm}{dt} = \alpha_m(1 - m) = \beta_m m,$$

$$\frac{dh}{dt} = \alpha_h(1 - h) = \beta_h h,$$

$$\frac{dn}{dt} = \alpha_n(1 - n) = \beta_n n,$$

for the solution that satisfies the initial conditions

$$V(0) = V_0,$$

$$m(0) = m_0,$$

$$h(0) = h_0,$$

$$n(0) = n_0,$$

where V_0 is the initial change in the membrane potential that is caused by the shock and m_0, h_0, n_0 are the resting values, that is, the values of $m(t), h(t), n(t)$, respectively, if V is the resting potential -70 mV. The agreement obtained is not entirely satisfactory. Certain discrepancies are discussed by Hodgkin and Huxley (1952d, pp. 525–526). In Hodgkin and Huxley (1952d, pp. 542–543) the extent to which these discrepancies can be attributed to known shortcomings in the equations (e.g., the too short delay in the rise of g_K) is discussed.

The numerical analysis reveals a good agreement between theoretical predictions and experimental observations of the absolute and relative refractory periods. There is good agreement for both the duration of the absolute refractory period and changes in $V(t)$ as the membrane returns from the refractory condition to the normal resting condition.

Another example of agreement between theoretical and experimental results is anode break excitation. In an experiment demonstrating anode break excitation, a current of negatively charged particles is made to flow inward through the membrane in such a way that the membrane potential is reduced from the resting value (-70 mV) to a value such as -100 mV. This current is continued for an interval of time significantly larger than τ_m, τ_h, or τ_n and the membrane potential is maintained at -100 mV during this interval. Then the current is suddenly stopped. The observed experimental result is that an action potential occurs. To obtain a theoretical description of anode break excitation, the Hodgkin–Huxley equations with $I = 0$ are solved for the solution with initial value:

$$V(0) = -100,$$
$$m(0) = m_0,$$
$$h(0) = h_0,$$
$$n(0) = n_0.$$

Good agreement is obtained between the experimental and theoretical description. The explanation of anode break excitation is that the decrease of the membrane potential decreases the potassium conductance and decreases sodium inactivation. Both of these effects persist when the current is stopped. Hence when the current is stopped and the membrane potential returns to the resting value, there is a reduced outward potassium current and an increased inward sodium current. The net flow of positive ions is inward and is sufficiently large so that depolarization occurs.

A striking example of agreement between theoretical and experimental results occurs in work with current clamps. A small constant current is passed through the membrane and the resulting changes of membrane potential measured. The current is supplied by an internal electrode so that the membrane is subject to a uniform current density. (Hence the name "current clamp.") Since the current is small, the changes that take place in V, m, h, n are small enough so that only the linear parts of the Hodgkin–Huxley equations need be considered. (The assumption that the linearized equations give a valid description is supported by experimental result that if the magnitude of the current is kept the same, but the direction or sign of the current changed, then the membrane potential also has the same magnitude but changes its sign.) The appropriate solutions of the linearized Hodgkin–Huxley equations agree very well with the data obtained from current-clamp experiments. This is a striking confirmation of the validity of the Hodgkin–Huxley equations because current-clamp experiments are very different from the voltage-clamp experiments that give rise to the Hodgkin–Huxley equations.

2.5.2.2 *Analysis of the full Hodgkin–Huxley equations*

So far we have been looking at examples of agreement between theory and experiment in cases where the potential is uniform on the axon, that is, the potential depends only on time and is independent of position on the axon. Now we wish to study situations in which the potential varies with position on the axon as well as time. This is what occurs in the normal physiological functioning of the nerve axon. The Hodgkin–Huxley equations as presented here cannot be used to study such situations because the equations describe conditions in which the membrane potential and the currents depend only on time. However, minor considerations from electricity theory make it possible to extend the equations so that they can be used to describe the situation in which the potential and the ionic conductances are functions of distance along the axon as well as time.

Following Jack, Noble, and Tsien (1975, p. 25ff) we obtain the extended model by combining the model of the space-clamp data and voltage-clamp data (i.e., the Hodgkin–Huxley equations) with a few statements from standard electrical theory.

First, from Ohm's law, it follows that if V is the membrane potential, R is the resistance per unit length of the interior of the nerve, i_a denotes flow of current in the axon, and x measures distance along the axon, then

$$\frac{\partial V}{\partial x} = -Ri_a. \tag{2.40}$$

Then if i_m denotes the total current across the membrane,

$$\frac{\partial i_a}{\partial x} = -i_m. \tag{2.41}$$

Differentiating (2.40) with respect to x and substituting from (2.41), we obtain

$$\frac{\partial^2 V}{\partial x^2} = R i_m. \tag{2.42}$$

[For a discussion of (2.40)–(2.42) see Jack, Noble, and Tsien (1975, p. 25ff).] But in the case of nerve conduction, we have, according to the Hodgkin–Huxley equations, that

$$i_m = C\frac{\partial V}{\partial t} + I_i$$

$$= C\frac{\partial V}{\partial t} + \bar{g}_{Na}m^3 h(V - V_{Na})S$$

$$+ \bar{g}_K n^4 (V - V_K)S + \bar{g}_l(V - V_l)S, \tag{2.43}$$

where C is the total capacitance of an axon membrane of unit length and S denotes the area of the axon membrane of unit length. Substituting from (2.43) into (2.42) we obtain the following equations that describe the changes in membrane potential and flow of ionic current across the membrane:

$$\frac{1}{R}\frac{\partial^2 V}{\partial x^2} = C\frac{\partial V}{\partial t} + \bar{g}_{Na}m^3 h(V - V_{Na})2\pi r$$

$$+ \left\{ \bar{g}_K n^4 (V - V_K) + \bar{g}_l(V - V_l) \right\}2\pi r,$$

$$\frac{\partial m}{\partial t} = \alpha_m(1 - m) - \beta_m m, \qquad (\mathcal{H}\text{-}\mathcal{H})$$

$$\frac{\partial h}{\partial t} = \alpha_h(1 - h) - \beta_h h,$$

$$\frac{\partial n}{\partial t} = \alpha_n(1 - n) - \beta_n n,$$

where r is the radius of the axon. We call these equations the full Hodgkin–Huxley equations. A solution of this system of partial differential equations consists of four functions $V(x, t)$, $m(x, t)$, $h(x, t)$, and $n(x, t)$ that satisfy the four partial differential equations. Now the general problem of solving the system ($\mathcal{H}\text{-}\mathcal{H}$) is extremely difficult. However, we are not interested in finding all the solutions of ($\mathcal{H}\text{-}\mathcal{H}$) but only those that would describe an

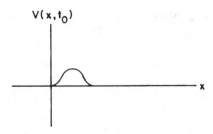

Figure 2.10.

action potential. An action potential is described by a function $V(x, t)$, which has the following property: At a given value $t = t_0$, the graph of $V(x, t_0)$ has the appearance shown in Fig. 2.10. At a later time $t = t_1 > t_0$, the graph of $V(x, t_1)$ has the same form as that of $V(x, t_0)$ except that it is moved to the right (see Fig. 2.11). Such a function can be written as

$$V(x, t) = W(x - \theta t),$$

where W is a function of one variable and θ is a positive constant that is the velocity with which the configuration moves to the right. This suggests that in order to obtain a description of the action potential we look for a solution of ($\mathcal{H}\text{-}\mathcal{H}$) of the form

$$(V(x - \theta t), m(x - \theta t), h(x - \theta t), n(x - \theta t)).$$

Such a solution is called a traveling wave solution and has long been a familiar concept in mathematical physics. The search for a traveling wave solution is much simpler than the search for a general solution of ($\mathcal{H}\text{-}\mathcal{H}$) because it can be reduced to the study of an ordinary differential equation. In order to carry out this reduction, we introduce the variable $\xi = x - \theta t$. Then by the familiar

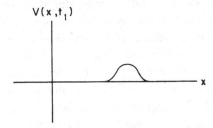

Figure 2.11.

chain rule from calculus, we have

$$\frac{\partial^2 V}{\partial x^2} = \frac{d^2 V}{d\xi^2},$$

$$\frac{\partial V}{\partial t} = -\theta \frac{dv}{d\xi},$$

$$\frac{\partial m}{\partial t} = -\theta \frac{dm}{d\xi},$$

$$\frac{dh}{\partial t} = -\theta \frac{dh}{d\xi},$$

$$\frac{\partial n}{\partial t} = -\theta \frac{dn}{d\xi},$$

and ($\mathscr{H}\text{-}\mathscr{H}$) becomes

$$\frac{1}{R} \frac{d^2 V}{d\xi^2} = C(-\theta) \frac{dV}{d\theta} + \left\{ \bar{g}_{\mathrm{Na}} m^3 h (V - V_{\mathrm{Na}}) \right\} 2\pi r$$

$$+ \left\{ \bar{g}_{\mathrm{K}} n^4 (V - V_{\mathrm{K}}) + \bar{g}_l (V - V_l) \right\} 2\pi r,$$

$$\frac{dm}{d\xi} = -\frac{1}{\theta} \left[\alpha_m (1 - m) - \beta_m m \right], \tag{2.44}$$

$$\frac{dh}{d\xi} = -\frac{1}{\theta} \left[\alpha_h (1 - h) - \beta_h h \right],$$

$$\frac{dn}{d\xi} = -\frac{1}{\theta} \left[\alpha_n (1 - n) - \beta_n n \right].$$

Thus the problem is reduced to the question of solving a system of ordinary differential equations: one second-order equation and three first-order equations. Usually, in order to solve (2.44) we must specify a value for θ, the velocity of the action potential. Also we are not searching for an arbitrary solution of (2.44). We are looking for a solution with the property that $V(\xi)$ has a configuration something like the graph in Fig. 2.10. Actually we impose a milder condition on $V(\xi)$: We require only that $\lim_{\xi \to \infty} V(\xi) = 0$.

Now we incorporate these two considerations into one problem, the solution of which will yield a theoretical estimate for the velocity of the action potential, that is, the following problem is studied: how to determine a value of θ for which the solutions $(V(\xi), m(\xi), h(\xi), n(\xi))$ of (2.44) are such that $\lim_{\xi \to \infty} V(\xi) = 0$. Calculations carried out by Hodgkin and Huxley (1952d, pp. 522–523) show that for a particular axon, the required conduction velocity was $\theta = 18.8$ m/s. The experimentally observed velocity for the same axon was 21.2 m/s. Thus the Hodgkin–Huxley equations yield a quite accurate estimate for the velocity of the action potential. This estimate is probably the most spectacular theoretical prediction of the Hodgkin–Huxley theory.

By solving (2.44) numerically with $\theta = 18.8$, the sodium and potassium currents during a propagated action potential can be computed and consequently the total addition of sodium and loss of potassium can be determined by integrating the corresponding ionic currents over the whole impulse. These theoretical results are in good agreement with experimentally determined values [see Hodgkin and Huxley (1952d, pp. 531–532)].

2.5.3 Drawbacks of the Hodgkin–Huxley equations

We have seen that numerical analysis of the Hodgkin–Huxley equations yields excellent agreement with a wide assortment of experimental results. Indeed when the arbitrariness of some of the steps in the derivation of the Hodgkin–Huxley equations is taken into account, the accurate theoretical predictions that come from analyzing them seem extraordinary.

Nevertheless, the Hodgkin–Huxley equations are not entirely accurate, and we describe now a couple of deficiencies or drawbacks, that is, we will describe two predictions made by the solutions of the equations that differ significantly from the experimental observations.

First, if I is fixed, then except for a narrow range of values of I, there is a periodic solution of the Hodgkin–Huxley equations [FitzHugh (1961)]. (Since all the solutions discussed here are approximate solutions obtained by numerical analysis, a "periodic solution" means an approximate solution that looks as if it is periodic.) But in the laboratory, stimulation of the squid giant axon by a step current produces only a finite train of up to four impulses [see Hagiwara and Oomura (1958) and Jakobsson and Guttman

(1980)]. Also in the laboratory, if I increases linearly with time, then if the rate of increase is below a certain threshold value, no action potential occurs. If the rate of increase of I is above that threshold value, one action potential occurs. But a numerical analysis of the Hodgkin–Huxley equations yields a periodic solution for all values of I except those in the interval mentioned before. If the phenomenon of accommodation is taken into account, it is possible to modify the Hodgkin–Huxley equations so that a closer agreement between experiment and theory can be obtained. Accommodation is a physiological process that produces a slow decay in a train of nerve impulses that result from a constant stimulation [see FitzHugh (1969)]. It can be regarded as an increase in the threshold of an excitable membrane when the membrane is subjected to a sustained subthreshold depolarizing stimulus or a stimulus that increases very slowly. A modification of the Hodgkin–Huxley equations that takes accommodation into account has been proposed by Adelman and FitzHugh (1975), and numerical analysis of the modified equations yields results that are in closer agreement with experiments using constant current stimulus.

It should be pointed out that the discussion of accommodation in Hodgkin and Huxley (1952d, pp. 537–538) is misleading because it suggests that the Hodgkin–Huxley equations can be used to describe accommodation. For a careful discussion of this point, see Jakobsson and Guttman (1980).

3

Nerve conduction:
Other mathematical models

3.1 Earlier models

We have presented the Hodgkin–Huxley equations first and in great detail because they constitute the most important model of nerve conduction. There are, however, other models that should be discussed. Here we will describe these models, their significance, and their status. However, as with the Hodgkin–Huxley equations, a detailed mathematical analysis will be postponed until Chapter 5.

In trying to understand the development of mathematical electrophysiology, it is enlightening to realize that a number of mathematical models, each consisting of a two-dimensional system of ordinary differential equations, were introduced in the 1930s. In these models, one of the dependent variables can be identified as $V(t)$, where $V(t)$ measures the potential difference across the membrane and the second variable, say $U(t)$, can be regarded as a recovery variable that tends to eliminate the excitability of the model after excitation has occurred and to bring about the end of the impulse. The models have the form

$$\frac{dV}{dt} = F(V, U),$$

$$\frac{dU}{dt} = G(V, U),$$

where F and G are well-behaved functions. It is no longer of interest to analyze these models in detail, but it is important to realize that such models exist and that part of Hodgkin and Huxley's accomplishment consisted in replacing rather general variables, such as the $U(t)$ in the description just given, with the concrete variables $m(t)$, $h(t)$, $n(t)$ that describe the sodium and potassium conductances. For an account of some of these models, see FitzHugh (1969).

3.2   The FitzHugh–Nagumo model

A second and very different model was introduced after the Hodgkin–Huxley equations. We have seen already that numerical analysis of the Hodgkin–Huxley equations showed that the solutions of the equations agreed, in the main, very well with experimental results. However, attempts to study the qualitative properties of solutions of the Hodgkin–Huxley equations proved fruitless partly because the equations are fundamentally nonlinear, partly because they are a four-dimensional system (whereas much of the known qualitative theory is applicable only to two-dimensional systems), and partly because the qualitative problems that arise in nerve conduction had not been studied earlier.

The general qualitative theory of differential equations originated from struggles to study the nonlinear differential equations that arise in celestial mechanics. It required the genius of Poincaré to shift the emphasis away from trying to obtain closed solutions (i.e., explicit formulas for solutions) and toward studying the general or qualitative behavior of solutions. For example, if it is known that the system of differential equations has a periodic solution and that all other nontrivial solutions approach this periodic solution, then a fairly complete understanding of the solutions of the system has been obtained even if no solution of the system has been explicitly computed. For mathematical models describing the axon, it is desirable to understand the qualitative behavior of solutions that reflects the threshold phenomenon and refractory periods. The Hodgkin–Huxley equations themselves, when first introduced, were too complicated to admit a qualitative analysis like this. Consequently, after the Hodgkin–Huxley equations had been introduced and studied numerically, efforts were made to obtain simpler systems of differential equations that would preserve the essential qualitative properties of the Hodgkin–Huxley equations. To do this successfully required an ingenious combination of mathematical and physiological reasoning. Notice that it could not be done simply on a mathematical basis because the essential qualitative properties of the Hodgkin–Huxley equations were not known. In fact, from the strictly mathematical viewpoint, the problem is impossible: One searches for a simpler system of differential equations that has the same essential qualitative properties as the Hodgkin–Huxley equations, but the qualitative properties of the

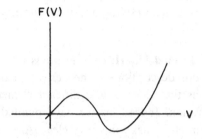

F(V)

V

Figure 3.1.

Hodgkin–Huxley equations are not at all well understood, and indeed the purpose of determining the simpler system is to aid in understanding the properties of the Hodgkin–Huxley equations! Thus from the purely logical viewpoint, finding such a simpler system is impossible. However, by taking into account the physiological background, FitzHugh (1961), and, independently, Nagumo, Arimoto, and Yoshizawa (1962), derived a two-dimensional system that is a desired simplification of the Hodgkin–Huxley equations. This system, usually called the FitzHugh–Nagumo equations, is the following two-dimensional system:

$$\frac{dV}{dt} = V - \frac{V^3}{3} - W + I,$$

$$\frac{dW}{dt} = \phi(V + a - bW),$$

where a, b, ϕ are positive constants, I denotes the membrane current, which is defined as any arbitrary function of time, V is the membrane potential, and W is a recovery variable. In a more general version of the FitzHugh–Nagumo equations, $-V + V^3/3$ is replaced by a function $F(V)$ that has the form indicated in Fig. 3.1. [see Rinzel (1976)].

3.3   **The Zeeman model**

The FitzHugh–Nagumo equations are a two-dimensional simplification of the four-dimensional Hodgkin–Huxley equations. A three-dimensional model has been proposed by Zeeman (1972, 1973, 1977). Zeeman derives his model by consideration of catastrophe theory and the use of the experimental data of Hodgkin and Huxley. Zeeman's model has serious deficiencies [these have been discussed in detail by Cronin (1981)] and it has no novel or useful properties.

3.4   Modifications of the Hodgkin–Huxley equations

It is clear from our description of the derivation of the Hodgkin–Huxley equations that there is a significant element of arbitrariness in the derivation. Indeed, as we have pointed out, Hodgkin and Huxley themselves make this very clear. Also it has been shown in Chapter 2 that the predictions made by the solutions of the Hodgkin–Huxley equations do not entirely agree with the experimental results. Consequently, it is reasonable to try to modify the Hodgkin–Huxley equations with a view to improving their powers of prediction and to making them more tractable to mathematical analysis. A number of modifications have been proposed, but no particular one has been widely accepted. The Hodgkin–Huxley equations themselves are still regarded as the most satisfactory mathematical model. We will indicate briefly a few of the modified models that have been proposed in order to indicate directions that future work may take; our descriptions are not to be regarded as an all-inclusive account. It should be pointed out that none of these modifications has withstood as thorough an examination as the Hodgkin–Huxley equations. In particular, no mathematical analysis of any of these models has ever been carried out to determine what action potential velocity they predict.

3.4.1   Modifications in the description of potassium and sodium conductances

Hodgkin and Huxley themselves point out a possible modification that would use a higher power of n in the expression for potassium conductance:

$$g_K = \bar{g}_K n^4$$

(see Chapter 2). Cole and Moore (1961) have proposed that for voltage clamps near the value of sodium equilibrium, a more accurate value for the exponent would be 25, although Cole (1975) later termed this a "tongue in cheek" suggestion. Other descriptions of the potassium conductances have been proposed by Tille (1965), Hoyt (1963), and FitzHugh (1965).

Modifications of the description of sodium conductance have also been suggested. Hoyt (1963, 1968) has proposed the use of a single second-order equation rather than a pair of coupled first-order equations. [Hodgkin and Huxley (1952d, p. 512) state that they chose to use a pair of coupled first-order equations because such equations were easier to apply to experimental results.] Hoyt and Adelman (1970) have shown that Hoyt's model gives a better description than the Hodgkin–Huxley equations of the experimental results for early changes of g_{Na} when an action potential occurs. Otherwise, the predictions made by the Hodgkin–Huxley model and the Hoyt model agree closely. Hoyt and Adelman draw from Hoyt's model conclusions concerning the physical mechanism. For a discussion of the validity of such conclusions, see Jakobsson (1973). Some other modifications are described briefly by Scott (1975, p. 505). Of particular interest is the brief discussion by Cole, FitzHugh, and Hoyt that follows the paper by FitzHugh (1965) in which all are in agreement that at that time there was no "best" model or set of equations.

3.4.2 The FitzHugh–Adelman model

In a more recent paper, Adelman and FitzHugh (1975) propose a modification of the Hodgkin–Huxley equations in which they take into account the potassium concentration in the periaxonal region and obtain, among other results, a more accurate description of accommodation during constant current stimulation. That is, the unrealistic periodic solutions described in Chapter 2 do not occur. Another modification has been proposed by Hodgkin (1975) and studied by Adrian (1975) to investigate the relation between sodium conductance and velocity of the nerve impulse. Huxley (1959) proposed a modification of the Hodgkin–Huxley equations that takes into account the effects of varying the calcium (Ca) ion concentration in the bath in which the squid axon is immersed. (Changes in the Ca ion concentration produce marked changes in the oscillatory properties of the membrane potential.) Huxley's model has been studied qualitatively by McDonough (1979).

3.4.3 The Hunter–McNaughton–Noble models

A very different line of study has been initiated by Hunter, McNaughton, and Noble (1975). One of their objectives is to develop models that closely mimic the Hodgkin–Huxley equations

(or the properties of real axons) but that are simple enough so that
closed form solutions that describe the propagation process can be
obtained. Having closed form solutions is obviously advantageous.
There is another advantage in studying models that are simpler
than the Hodgkin–Huxley equations. The full voltage-clamp analysis that is the basis for the derivation of the Hodgkin–Huxley
equations is only possible for certain excitable cells or for a limited
range of conditions. For example, as we will see later, experimental
study of cardiac muscle is far more difficult than the experimental
study of the squid axon. In such cases, it becomes necessary to
resort to simpler models in order to study some excitation and
conduction processes in a quantitative manner.

The work of Hunter et al. (1975) also includes a more general
approach to the study of sodium conductivity. This more general
approach is an extension of the approach used in Hoyt's model
(1963), but it is based on later experimental results. Experimental
work by Goldman and Schauf (1972) suggests that there may be a
relationship between activation and inactivation. This, in turn,
suggests that the sodium conductance g_{Na} should be described
differently. In the Hodgkin–Huxley theory,

$$g_{Na} = \bar{g}_{Na} m^3 h,$$

where m and h are activation and inactivation variables governed
by the equations

$$\frac{dm}{dt} = \alpha_m(V)m - [\beta_m(V)](1-m),$$

$$\frac{dh}{dt} = \alpha_h(V)h - [\beta_h(V)](1-h).$$

Hunter et al. (1975) propose to describe g_{Na} in terms of a linear
nth-order equation

$$g_{Na} + \left(\frac{1}{\alpha_1} + \frac{1}{\alpha_2} + \cdots + \frac{1}{\alpha_n}\right)\frac{dg_{Na}}{dt}$$

$$+ \left(\frac{1}{\alpha_1\alpha_2} + \frac{1}{\alpha_1\alpha_3} + \cdots + \frac{1}{\alpha_1\alpha_n}\right)\frac{d^2 g_{Na}}{dt}$$

$$+ \cdots + \left(\frac{1}{\alpha_1\alpha_2,\ldots,\alpha_n}\right)\frac{d^n g_{Na}}{dt^n} = (g_{Na})_\infty,$$

where $\alpha_1, \alpha_2, \ldots, \alpha_n$ and (g_{Na}) are functions of V.

3.5   The Lecar–Nossal stochastic model

Finally, we describe briefly a very important modification of the Hodgkin–Huxley theory, which will be studied in more detail in Chapter 6.

Since the functions that appear in the Hodgkin–Huxley equations are continuously differentiable functions, the standard existence and uniqueness theorems for solutions of the differential equations are applicable. Hence a solution that satisfies the initial conditions $V(0) = V_0$, $m = m_\infty(0)$, $h = h_\infty(0)$, and $n = n_\infty(0)$ is unique. Thus if V_0 is the threshold value, the prediction given by the differential equation is unique: An action potential always occurs, that is, the nerve always fires. But this prediction does not agree with experimental observations. In experiments, if $V(0)$ is set at the threshold value, the axon does not always respond by firing. Firing occurs only for a certain proportion of the number of times that stimulus occurs. The reason for the varying responses of the real axon to threshold stimulus is thought to be the random fluctuations or "noise" in the resting potential. This suggests that the Hodgkin–Huxley equations would give a more realistic description of the behavior of the axon if they contained a random or stochastic term. Also the study of the Hodgkin–Huxley equations modified by the addition of such a stochastic term would make a theoretical connection between probabilistic firing of axons and electrical noise generated across the nerve membrane possible. It should be emphasized at this point that the addition of a stochastic term is not just for the purpose of constructing a novel mathematical exercise. Random firing of axons exposed to near-threshold stimuli has been studied by physiologists for many years. It is a serious subject in physiology and is deserving of a quantitative description.

A study in this direction has been carried out by Lecar and Nossal (1971). They used the FitzHugh–Nagumo equations and included a description of the noisy nerve by adding a two-dimensional Brownian motion. The description given by this model is not only more realistic in that it includes a description of the noise, but it turns out also to give a better description of the threshold than the FitzHugh–Nagumo equations give. We will discuss these advantages in Chapter 6.

4

Models of other electrically excitable cells

4.1 Introduction

The techniques developed by Hodgkin and Huxley for the study of the squid axon have been applied in the ensuing years by many other researchers. In a certain sense, the techniques remain the same, that is, voltage-clamp experiments followed by a quantitative analysis in which activation and inactivation variables are introduced and then described by differential equations. But there are many obstacles to these applications of Hodgkin and Huxley's techniques: The analysis of the ionic current becomes much more complicated because the descriptions of some of the currents are more intricate and because the number of distinct components of the current is in some cases much larger than for the squid axon. (Later we shall describe a mathematical model for the cardiac Purkinje fiber in which the ionic current has nine components.) Also the use of voltage-clamp methods is much more difficult in some cases. For example, voltage-clamp techniques were used successfully in the study of cardiac fibers for the first time in 1964, and the voltage-clamp technique used in the study of striated muscle fibers was not developed until the late 1960s.

The purpose of this chapter is to describe mathematical models (systems of nonlinear ordinary differential equations) of a number of electrically excitable cells that can be investigated by using Hodgkin–Huxley techniques. Since our primary concern is the derivation and study of these mathematical models that stem from the experimental studies, it is easy to forget or lose sight of the extensive and taxing work that goes into successful experiments. In all of this discussion, it should be kept in mind that underlying these theoretical considerations is a great deal of original, ingenious, and difficult experimental work.

Besides their intrinsic interest, the models in this chapter are testimony to the remarkable insight and accomplishments of Hodgkin and Huxley. As we shall see, these models and the physiological systems they describe differ in many ways from the Hodgkin–Huxley equations and the squid axon that they describe. Nevertheless these models are based on the same viewpoint as that developed by Hodgkin and Huxley; part of their accomplishment was the development of this remarkably flexible mode of description that can be adapted to many different physiological systems.

4.2 The myelinated nerve fiber

From the biological viewpoint, it is clear that the speed with which a nerve impulse travels is crucially important to the animal, and we have already seen in Chapter 2 how a theoretical estimate of the speed of the nerve impulse was obtained from the Hodgkin–Huxley equations. The theoretical speed depends on the square root of the radius r of the cross section of the axon. In order to see this, we return to the first of the equations in system ($\mathscr{H}$-$\mathscr{H}$) in Chapter 2. If we consider only solutions of the form $V(x - \theta t)$, then we have at once

$$\frac{\partial^2 V}{\partial x^2} = \theta^{-2} \frac{\partial^2 V}{\partial t^2}$$

and the first equation in system ($\mathscr{H}$-$\mathscr{H}$) becomes

$$\frac{1}{R\theta^2} \frac{\partial^2 V}{\partial t^2} = C \frac{\partial V}{\partial t} + 2\pi r l \{I_i\}, \tag{4.1}$$

where R is the resistance per unit length of the axon, θ is the speed of the traveling wave that describes the nerve impulse, C is the capacitance of the membrane enclosing a unit length of axon, and I_i is the ionic current flow per unit area through the membrane. Now if R_i is the resistance per unit length of an axon of unit cross section (the specific resistivity of the axoplasm), then

$$R = \frac{R_i}{\pi r^2}.$$

If C_M is the capacitance per unit area of the membrane, then

$$C = 2\pi r C_M.$$

Then (4.1) becomes

$$\frac{\pi r^2}{R_i \theta^2} \frac{\partial^2 V}{\partial t^2} = 2\pi r C_M \frac{\partial V}{\partial t} + 2\pi r I_i$$

or

$$\frac{r}{2 R_i \theta^2} \frac{\partial^2 V}{\partial t^2} = C_M \frac{\partial V}{\partial t} + I_i.$$

Since R_i, C_M, and I_i are independent of r, and since $\partial V/\partial t$ and d^2V/dt^2 are independent of r, then r/θ^2 must be independent of r and hence

$$\theta = k\sqrt{r},$$

where k is a proportionality constant. This result suggests that in a nerve axon with the structure of the squid axon, it would be necessary to quadruple the radius of the cross section of the axon in order to double the speed of the nerve impulse.

As pointed out earlier, the radius of the squid axon is unusually large. Probably the chief reason for this is that the impulse that travels along the squid giant axon is an impulse that orders, "Flee," and consequently speed of transmission is particularly important. The unusually large radius of the squid axon permits a comparatively high speed of transmission.

However, for most animal species, the direction of evolution has been not toward nerve axons of large radius, but instead toward the development of myelinated axons in which a different kind of transmission of impulse takes place. As described in Chapter 2, the myelinated axon is surrounded by a sheath of fatty material called myelin. The sheath is interrupted at intervals of about 1 mm by short gaps called nodes of Ranvier. The myelin layer is a fairly good insulator and little current crosses the membrane except at the nodes, where excitation occurs. During impulse conduction, the excitation jumps from node to node. This mode of transmission permits much higher speed of transmission even though the radius of the axon remains small. [For a detailed analysis of this, see Scott (1975), especially p. 516.] Many nerve axons are myelinated; for example, all the nerve fibers, except the smallest ones, in vertebrates are myelinated.

The experimental study of the myelinated axon consists in carrying out voltage-clamp experiments on the nodal membrane, that is, a section of the axon on which there is no sheath of myelin. A long series of experiments on a particular myelinated nerve axon, the sciatic nerve of the clawed toad (*Xenopus laevis*), was carried out by Frankenhaeuser and several co-workers. They performed voltage-clamp experiments and derived a mathematical model analogous to the Hodgkin–Huxley equations. Finally, Frankenhaeuser and Huxley (1964) used numerical methods to obtain approximate solutions of this mathematical model and compared these numerical results with experimentally recorded action potentials. Complete references to the earlier work are given by Frankenhaeuser and Huxley (1964).

The results show that there are marked resemblances between the ionic currents that are obtained when a voltage clamp is applied to a squid axon and when a voltage clamp is applied to the nodal membrane of the myelinated axon, but there are also important differences: for example, the relationship between ionic current and membrane potential is more complicated for the nodal membrane than for the squid axon. Hence the simple description of I_{Na} used by Hodgkin and Huxley, that is, the equation

$$I_{Na} = g_{Na}(V - V_{Na}),$$

cannot be used. Instead, the description of I_{Na} is based upon the constant field equation due to Goldman (1943/44) and described by Hodgkin and Katz (1949). The constant field equation takes into account the effect of the ion concentrations in the membrane. (The name "constant field equation" stems from the fact that, in the derivation of the equation, one of the basic assumptions is that the electric field can be regarded as constant throughout the membrane.) From the constant field equation, the following expression is obtained for I_{Na}:

$$I_{Na} = P_{Na}\frac{F^2 E}{RT}[Na]_o\frac{[\exp\{(E - E_{Na})F/RT\} - 1]}{\exp[EF/RT] - 1},$$

$$(4.2)$$

where E is the value of the membrane potential at the time of measurement of current, E_{Na} is the sodium equilibrium potential,

$[Na]_o$ is the external sodium concentration, R the gas constant, T the absolute temperature, F the Faraday constant, and P_{Na} is the sodium permeability, which in this work plays a role similar to g_{Na} in the Hodgkin–Huxley equations.

Also, while the ionic current in the squid axon has three components–the sodium current I_{Na}, the potassium current I_K, and the leakage current I_L (observe the change in notation from the Hodgkin–Huxley notation for leakage current)–the ionic current across the nodal membrane has four components: an initial current I_{Na} carried mainly by sodium ions; a delayed current I_K carried by potassium ions; a second "nonspecific" delayed current I_P which is carried to a large extent by sodium ions, but possibly also by other ions such as calcium; and a leakage current I_L carried by both sodium and potassium. The currents are described as follows:

$$I_L = g_L(V - V_L),$$

where $g_L = 30.3$ mΩ^{-1}/cm^2, V denotes displacement of the membrane potential from the resting potential (i.e., $V = E - E_r$, where E_r is the resting potential that is taken to be -70 mV), and $V_L = 0.026$ mV. The numerical value V_L is chosen so that $V = 0$ if no external current is applied to the axon. It is necessary to allow this adjustment for V_L because the data for I_{Na}, I_K, and I_P, which we will shortly write, are made up entirely from experimental data; at $V = 0$, it is necessary that

$$I_i = I_{Na} + I_K + I_P + I_L = 0.$$

Now according to the Nernst formula

$$E_{Na} = \frac{RT}{F} \ln \frac{[Na]_o}{[Na]_i}$$

or

$$-\frac{FE_{Na}}{RT} = -\ln \frac{[Na]_o}{[Na]_i}$$

or

$$\exp\left\{-\frac{FE_{Na}}{RT}\right\} = \frac{[Na]_i}{[Na]_o}.$$

Substituting in (4.2), we have

$$I_{Na} = P_{Na} \frac{F^2 E}{RT} [Na]_o \frac{(\exp(EF/RT))([Na]_i/[Na]_o) - 1}{\exp(EF/RT) - 1}$$

$$= P_{Na} \frac{F^2 E}{RT} \frac{[Na]_o - [Na]_i \exp(EF/RT)}{1 - \exp(EF/RT)}.$$

In this case, the outside sodium concentration $[Na]_o$ is 114.5 mM, the inside sodium concentration $[Na]_i$ is 13.74 mM, and

$$P_{Na} = \overline{P}_{Na} m^2 h,$$

where $\overline{P}_{Na}$ is the sodium permeability constant that has the value 8×10^{-3} cm/s, and m and h are the sodium activation and inactivation constants, respectively, that satisfy the equations

$$\frac{dm}{dt} = \alpha_m (1 - m) - \beta_m m,$$

$$\frac{dh}{dt} = \alpha_h (1 - m) - \beta_h h.$$

(See Table 4.1 for the rate coefficients $\alpha_m, \alpha_h, \beta_m, \beta_h$. The constants A, B, C that appear in Table 4.1 are listed in Table 4.2.) Similarly,

$$I_K = P_K \frac{EF^2}{RT} \frac{[K]_o - [K]_i \exp(EF/RT)}{1 - \exp(EF/RT)},$$

where $[K]_o$ is the outside potassium concentration, which is 2.5 mM, $[K]_i$ is the inside potassium concentration, which is 120 mM, and

$$P_K = P_K' n^2,$$

where P_K' is the potassium permeability constant, which is 1.2×10^{-3} cm/s, and

$$\frac{dn}{dt} = \alpha_n (1 - n) - \beta_n n.$$

(See Table 4.1 for α_n and β_n and Table 4.2 for the corresponding constants A, B, C.) Finally

$$I_p = P_p \frac{EF^2}{RT} \left[\frac{[Na]_o - [Na]_i \exp(EF/RT)}{1 - \exp(EF/RT)} \right],$$

Table 4.1. *Rate coefficient functions*

$$\alpha_h(V) = \frac{A(B-V)}{\left\{1 - \exp\left(\dfrac{V-B}{C}\right)\right\}} \qquad \alpha_n(V) = \frac{A(V-B)}{\left\{1 - \exp\left(\dfrac{B-V}{C}\right)\right\}}$$

$$\beta_h(V) = \frac{A}{\left\{1 + \exp\left(\dfrac{B-V}{C}\right)\right\}} \qquad \beta_n(V) = \frac{A(B-V)}{\left\{1 - \exp\left(\dfrac{V-B}{C}\right)\right\}}$$

$$\alpha_m(V) = \frac{A(V-B)}{\left\{1 - \exp\left(\dfrac{B-V}{C}\right)\right\}} \qquad \alpha_p(V) = \frac{A(V-B)}{\left\{1 - \exp\left(\dfrac{B-V}{C}\right)\right\}}$$

$$\beta_m(V) = \frac{A(B-V)}{\left\{1 - \exp\left(\dfrac{V-B}{C}\right)\right\}} \qquad \beta_p(V) = \frac{A(B-V)}{\left\{1 - \exp\left(\dfrac{V-B}{C}\right)\right\}}$$

where

$$P_p = \overline{P}_p p^2,$$

and $\overline{P}_p$ is the nonspecific permeability constant and

$$\frac{dp}{dt} = \alpha_p(1-p) - \beta_p p.$$

(See Table 4.1 for α_p and β_p.)

Table 4.2. *Values of constants A, B, C associated with α and β*

	A (ms^{-1})	B (mV)	C (mV)	Reference
α_h	0.1[a]	-10[a]	6[a]	Frankenhaeuser (1960)
β_h	4.5	$+45$	10	Frankenhaeuser (1963b)
α_m	0.36	$+22$	3	Frankenhaeuser (1960)
β_m	0.4	$+13$	20	
α_p	0.006	$+40$	10	Frankenhaeuser (1963a)
β_p	0.09	-25	20	
α_n	0.02	$+35$	10	
β_n	0.05	$+10$	10	

[a] Values modified to give a single curve for experimental results in Frankenhaeuser (1960, Fig. 4).

The equation for dV/dt is

$$\frac{dV}{dt} = \frac{1}{C_m}\{I - (I_{Na} + I_K + I_P + I_L)\},$$

where C_m is the membrane capacitance and I is the total current. Thus the mathematical model for the myelinated nerve fiber has the form

$$\frac{dV}{dt} = \frac{1}{C_m}\left\{ I - \overline{P}_{Na}m^2h\frac{EF^2}{RT}\left[\frac{[Na]_o - [Na]_i\exp(EF/RT)}{1 - \exp(EF/RT)}\right]\right.$$

$$- P'_K n^2 \frac{EF^2}{RT}\frac{[K]_o - [K]_i\exp(EF/RT)}{1 - \exp(EF/RT)}$$

$$\left. - \overline{P}_p p^2 \frac{EF^2}{RT}\frac{[Na]_o - [Na]_i\exp(EF/RT)}{1 - \exp(EF/RT)} - g_L(V - V_L)\right\},$$

$$\frac{dy}{dt} = \alpha_y(1 - y) - \beta_y y, \qquad y = m, h, n, p.$$

It is shown by Frankenhaeuser and Huxley (1964) that a numerical solution of this system that describes an action potential agrees satisfactorily with experimentally recorded action potentials. Details of the comparison between the theoretical prediction and the experimental results are given in Frankenhaeuser and Huxley (1964).

The mathematical problems that must be studied for this model are basically the same problems that must be studied for the Hodgkin–Huxley equations. Again it seems reasonable to expect that there should be a unique equilibrium point that is globally asymptotically stable. Also, it would be desirable to obtain a qualitative analysis that shows that the solutions display threshold phenomena and a refractory period.

4.3 Striated muscle fiber

Techniques for making experimental voltage-clamp studies of a striated muscle fiber were not developed until the late 1960s and a mathematical model was derived in 1970 by Adrian, Chandler, and Hodgkin (1970). In certain respects, the mathematical model derived is similar to the Hodgkin–Huxley equations. There are three components of the ionic current, all governed quantitatively in a way similar to the three components in the Hodgkin–Huxley model. However, the variation of V'_K (the equilibrium potential corresponding to the delayed potassium current) with external potassium concentration suggests that an equivalent circuit description of the membrane more complicated than the Hodgkin–Huxley membrane description should be used. As a result, the description includes the variation of two potential differences.

This model is not entirely satisfactory and we include a description of it to indicate a direction of study in physiology rather than for the value of the model itself. An improvement of this model that uses a more elaborate equivalent circuit description of the membrane has been given by Adrian and Peachey (1973).

As with the squid axon, there are three components of the ionic current: sodium current I_{Na}, potassium current I_K, and leakage current I_L. These ionic currents are described by

$$I_{Na} = \bar{g}_{Na} m^3 h (V - V'_{Na}),$$

$$I_K = \bar{g}_K n^4 (V - V'_K),$$

$$I_L = g_L (V - V_L),$$

where

$$V'_{Na} = \frac{RT}{F} \ln\left[\frac{[Na]_o + b[K]_o}{[Na]_i + b[K]_i} \right],$$

in which $[Na]_o$, $[K]_o$, $[Na]_i$, and $[K]_i$ have the usual meanings (we will not give the numerical values here) and the constant b has the value $1/12$. The reason that the usual sodium equilibrium potential V_{Na} is not used is that the muscle fiber is immersed in a fluid (Ringer fluid containing 350 mM sucrose), the purpose of which is to reduce the mechanical contractions that would normally be performed by a muscle fiber in the range of depolarizations considered in the experiments. The value used for V'_{Na} is 50 mV. V'_K is defined in a similar way and the value used for V'_K is -70 mV. The other constants are $V_L = -100$ mV, $\bar{g}_L = 0.55$ mΩ^{-1}/cm^2, $\bar{g}_{Na} = 61$ mΩ^{-1}/cm^2, and $\bar{g}_K = 9.8$ mΩ^{-1}/cm^2. The activation and inactivation variables m, h, and n each satisfy a differential equation of

Table 4.3. *The coefficients* α_y *and* β_y

Muscle	Squid giant axon
$\alpha_m = \dfrac{\bar{\alpha}_m(V - \bar{V}_m)}{1 - \exp - \dfrac{(V - \bar{V}_m)}{10}}$	$\alpha_m = \dfrac{\bar{\alpha}_m(V - \bar{V}_m)}{1 - \exp - \dfrac{(V - \bar{V}_m)}{10}}$
$\beta_m = \bar{\beta}_m \exp - \dfrac{(V - \bar{V}_m)}{18}$	$\beta_m = \bar{\beta}_m \exp - \dfrac{(V - \bar{V}_m)}{18}$
$\alpha_h = \bar{\alpha}_h \exp - \dfrac{(V - \bar{V}_h)}{14.7}$	$\alpha_h = \bar{\alpha}_h \exp - \dfrac{(V - \bar{V}_h)}{20}$
$\beta_h = \bar{\beta}_h \left[1 + \exp - \dfrac{(V - \bar{V}_h)}{7.6}\right]^{-1}$	$\beta_h = \bar{\beta}_h \left[1 + \exp - \dfrac{(V - \bar{V}_h)}{10}\right]^{-1}$
$\alpha_n = \dfrac{\bar{\alpha}_n(V - \bar{V}_n)}{1 - \exp - \dfrac{(V - \bar{V}_n)}{7}}$	$\alpha_n = \dfrac{\bar{\alpha}_n(V - \bar{V}_n)}{1 - \exp - \dfrac{(V - \bar{V}_n)}{10}}$
$\beta_n = \bar{\beta}_n \exp - \dfrac{(V - \bar{V}_n)}{40}$	$\beta_n = \bar{\beta}_n \exp - \dfrac{(V - \bar{V}_n)}{80}$

the form

$$\frac{dy}{dt} = \alpha_y(1 - y) - \beta_y y, \qquad y = m, h, n,$$

where the equation for m is the same as that in the Hodgkin–Huxley equations for the squid axon and the equations for h and n have similar forms. The coefficients α_y and β_y ($y = m, h, n$) are given in Table 4.3. In the table

$$\bar{\alpha}_m = 0.1, \qquad \bar{\beta}_m = 4,$$

$$\bar{\alpha}_h = 0.07, \qquad \bar{\beta}_h = 1,$$

$$\bar{\alpha}_n = 0.01, \qquad \bar{\beta}_n = 0.125.$$

The equivalent circuit chosen to describe an element of membrane in the striated muscle fiber is sketched in the accompanying diagram.

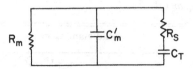

The values chosen are $C'_m = 1 \; \mu\text{F}/\text{cm}^2$, $C_T = 4 \; \mu\text{F}/\text{cm}^2$, and $R_S = 150 \; \Omega/\text{cm}^2$. The value of R_m depends on the external solution. Values are given by Adrian et al. (1970, p. 615). Let V_T denote the potential across capacitance C_T. The remaining differential equations are

$$\frac{dV}{dt} = -\frac{I_i}{C'_M} - \frac{V - V_T}{R_S C'_M}$$

or

$$\frac{dV}{dt} = -\frac{1}{C'_M} \left[g_{\text{Na}} m^3 h (V - V'_{\text{Na}}) + g_{\text{K}} n^4 (V - V'_{\text{K}}) \right.$$

$$\left. + g_L (V - V_L) \right] - \frac{V - V_T}{R_S C'_M} \quad (4.3)$$

and

$$\frac{dV_T}{dt} = \frac{V - V_T}{R_S C_T}. \quad (4.4)$$

Thus the entire mathematical model consists of (4.3), (4.4) and the equations

$$\frac{dy}{dt} = \alpha_y (1 - y) - \beta_y y, \qquad y = m, h, n,$$

which were described earlier.

On the basis of these equations, propagated action potentials and conduction velocities were computed by Adrian et al. The computed results were found to be in "reasonable agreement" with experimentally observed values.

A mathematical study of this model or a refinement of it would consist of studying the same kind of problems that are important in the study of the model of the squid axon and the model of the myelinated nerve.

4.4 Cardiac fibers

We shall describe mathematical models for two kinds of electrically excitable cardiac cells: the cardiac Purkinje fiber and the ventricular myocardial fiber. Developing voltage-clamp methods for the study of cardiac fibers proved to be a difficult task and models based on experimental results were not developed until the 1970s, and consequently realistic mathematical models appeared only fairly recently.

4.4.1 The cardiac Purkinje fiber

4.4.1.1 Physiological functions of the Purkinje fiber

Our first step is to briefly describe the physiological functions of the cardiac Purkinje fiber. The initiation of the electrical activity that governs the heartbeat is carried out by a pacemaker mechanism that is located at a position in the heart called the sino-atrial node (SA node). (The SA node is not the only region of the heat that possesses a pacemaker mechanism, but its rate of beating is higher than that of any other pacemaker region and so it sets the pace of the heart as a whole.) The electrical impulses generated by the pacemaker mechanism are conducted to the various parts of the heart, first to the atria, then through a small strip of tissue called the atrioventricular (AV) node, and finally to the ventricles. The ventricles are the two large chambers of the heart into which the blood is brought from the two smaller thin-walled chambers called the atria and from which the blood is pumped to lung circulation and systemic circulation. The electrical impulse is conducted comparatively slowly through the AV node. Then it is conducted more rapidly to all of the walls of the ventricles. The electrical impulse causes the muscles in the ventricular walls to contract, thus producing the heartbeat and pumping the blood into the circulatory system. Consequently, it is important that the muscles in the ventricular walls contract almost simultaneously so that the blood is forced into the arteries at high pressure. Thus, in turn, it is important that the electrical impulse be conducted very rapidly from the AV node to the ventricular walls.

The electrical impulses are conducted from the AV node to the ventricular walls by the cardiac Purkinje fibers and their large size insures that the conduction is rapid.

Conduction of electrical impulses that originate in the pacemaker region is the primary function of the cardiac Purkinje fibers, but they also have a secondary function. The Purkinje fibers fire spontaneously at a regular rate. This regular rate is slower than the rate of firing of the pacemaker mechanism and, consequently, under normal conditions (i.e., when the Purkinje fiber conducts an electrical impulse that originates from the pacemaker mechanism) the Purkinje fiber does not fire spontaneously. However, if the electrical impulses from the pacemaker mechanism do not reach the Purkinje fibers (this happens, for example, in a clinical condition called AV block), then the spontaneous firing of the Purkinje fibers plays a major role. The Purkinje fibers then act as a pacemaker and the heart beats at the rate of the spontaneous firing of the Purkinje fibers. Thus a secondary function of the Purkinje fibers is to act as a "backup" pacemaker.

4.4.1.2 *The Noble model of the cardiac Purkinje fiber*

Before voltage-clamp techniques were completely developed for the study of cardiac Purkinje fibers, Noble (1962) proposed a modification of the Hodgkin–Huxley equations that can be used to describe the action potentials and pacemaker potentials of cardiac Purkinje fibers. That is, solutions of the system of differential equations proposed by Noble describe fairly accurately these action potentials and pacemaker potentials. More realistic models, based on voltage-clamp data were derived later by McAllister, Noble, and Tsien (1975) and Di Francesco and Noble (1985), but the earliest model developed by Noble is worth studying first because, like the Hodgkin–Huxley model, it is a four-dimensional system and there is the possibility of using analysis that was developed to study the Hodgkin–Huxley equations to analyze this new model. (The McAllister–Noble–Tsien mode is a 10-dimensional model and, at present, no mathematical analysis beyond numerical analysis has been made of that model.)

The Purkinje fiber differs from the squid axon in that depolarization decreases the potassium permeability of the membrane. During large depolarizations, part of this decrease seems to be transient and the potassium permeability then slowly increases. To account for this, Noble assumed that the potassium ions move through two types of channel in the membrane. [This assumption was based on experimental measurements made by Noble and his co-workers; see Noble (1979, p. 21).] In one channel, the potassium conductance (g_{K_1}) is assumed to be an instantaneous function of the membrane potential and decreases when the membrane is depolarized. In the other channel, the conductance (g_{K_2}) slowly rises when the membrane is depolarized. The empirical expression for g_{K_1} is:

$$g_{K_1} = 1.2 \exp\left[\frac{(-E - 90)}{50} \right] + 0.05 \exp\left[\frac{E + 90}{60} \right],$$

where potential E is the membrane potential, that is, "inside minus outside." (Note that the Hodgkin–Huxley equations concern the quantity V, i.e., the deviation of the membrane potential from the resting potential, whereas in this work, the quantity E, i.e., the membrane potential itself, is studied.) The term g_{K_2} is described by

$$g_{K_2} = 1.2 n^4,$$

where

$$\frac{dn}{dt} = \alpha_n (1 - n) - \beta_n n,$$

$$\alpha_n = \frac{0.0001(-E - 50)}{\exp[(-E - 50)/10] - 1},$$

$$\beta_n = 0.002 \exp\left(\frac{-E - 90}{80} \right).$$

The sodium conductance g_{Na} is described by equations similar to those used to describe the sodium current in the squid axon. That is,

$$g_{Na} = m^3 h \bar{g}_{Na},$$

where $\bar{g}_{Na}$ is a positive constant and m, h satisfy the equations

$$\frac{dm}{dt} = \alpha_m (1 - m) - \beta_m m,$$

$$\frac{dh}{dt} = \alpha_h (1 - h) - \beta_h h,$$

where

$$\alpha_m = \frac{0.1(-E-48)}{\exp[(-E-48)/15]-1},$$

$$\beta_m = \frac{0.12(E+8)}{\exp[(E+8)/5]-1},$$

$$\alpha_h = 0.17\exp\left[\frac{(-E-90)}{20}\right],$$

$$\beta_h = \left\{\exp\left(\frac{-E-42}{10}\right)+1\right\}^{-1}.$$

In addition to the sodium and potassium currents, a leakage current, consisting at least in part of chloride ions, is assumed to exist. Denoting this current by I_{An} (An stands for anion) and denoting the anion equilibrium potential and the anion conductance by g_{An}, we have

$$g_{An} = \frac{I_{An}}{E-E_{An}}$$

(g_{An} is regarded as a constant, but various values of g_{An} are used in calculations in order to reproduce the effects of anions of different permeabilities.) Thus the total membrane current is

$$I = C_m\frac{dE}{dt} + I_{Na} + I_K + I_{An}$$

or

$$I = C_m\frac{dE}{dt} + g_{Na}(E-E_{Na})$$
$$+ g_K(E-E_K) + g_{An}(E-E_{An}),$$

where C_m is the membrane capacitance and is taken to be 12 $\mu F/cm^2$ and $g_K = g_{K_1} + g_{K_2}$. As in the Hodgkin–Huxley description of the squid axon, this description is space-clamped, that is, E at a given time is assumed to have the same value at all points along the fiber. Consequently there is no current flow along the fiber and, therefore, unless a current is applied, the total membrane current is 0. Thus, for the complete system of equations, we obtain

$$\frac{dE}{dt} = -\frac{1}{C_m}\left\{ m^3 h \bar{g}_{Na}(E - E_{Na}) + \left(g_{K_1} + 1.2n^4\right)\right.$$

$$\left. \times (E - E_K) + g_{An}(E - E_{An})\right\}, \quad \text{(N1)}$$

$$\frac{dm}{dt} = \alpha_m(1 - m) - \beta_m m, \qquad \text{(N2)}$$

$$\frac{dh}{dt} = \alpha_h(1 - h) - \beta_h h, \qquad \text{(N3)}$$

$$\frac{dn}{dt} = \alpha_n(1 - n) - \beta_n n. \qquad \text{(N4)}$$

There is a detailed discussion in Noble (1962) of the similarities and discrepancies between the experimental results and the numerical solutions of the system consisting of (N1)–(N4). In certain respects these equations closely resemble the Hodgkin–Huxley equations, but the solutions can be expected to exhibit quite different behavior. For example, as pointed out in Chapter 2, solutions of the Hodgkin–Huxley equations would generally be expected to approach an equilibrium point because after an action potential is formed, the membrane potential returns to the resting value. But the Purkinje fibers exhibit rhythmic electrical behavior. Hence, it is reasonable to expect to find periodic solutions of Noble's system of equations, and numerical studies suggest [see, e.g., Noble (1962, p. 329)] that there exist such periodic solutions.

4.4.1.3 *The McAllister–Noble–Tsien model of the cardiac Purkinje fiber*

A mathematical model of the cardiac Purkinje fiber based on voltage-clamp experiments was derived by McAllister, Noble, and Tsien (1975). This model is a more realistic and accurate description of the cardiac Purkinje fiber than the Noble model. (A detailed comparison of the two models is given by McAllister et al.) But it does have certain drawbacks. First, it is based on a "mosaic of experimental results" [McAllister et al. (1975, page 1)], that is, unlike the Hodgkin–Huxley model of the squid, the required information for the model cannot be obtained from voltage-clamp analysis of a single fiber. Also, the experimental results (especially the voltage-clamp data) are incomplete in some ways, and the model has one important deficiency: the sodium current is inadequate to fully account for the upstroke velocity without making further assumptions [see McAllister et al. (1975, pp. 53–54)]. Nevertheless, it is useful and enlightening to study this model: However tentative it is in the view of physiologists, it indicates an important direction of work in the mathematics of physiology, and it conveys to the mathematician some idea of the complexity and difficulty of the problems in this subject.

The McAllister–Noble–Tsien (MNT) model is like the Noble model in that it is basically a quantitative description of ionic currents across the membrane. But later experimental results have made possible a more detailed analysis of these currents, and so the model is considerably more complicated.

First we describe the various components of the ionic current. As in the Noble model, we let E denote the membrane potential and E_{Na} denote the sodium equilibrium potential, and use similar notation for the other equilibrium potentials. There are two time-dependent inward currents: i_{Na} and i_{si}. (The term "time-dependent" means that the current i_{Na} is not simply a function of E. Later the term "time-independent" will be used. A time-independent current is a current that depends only on E. Since E itself may be a function of time, this terminology is not entirely logical.) The current i_{Na} resembles the squid sodium current and can be represented as

$$i_{Na} = \bar{i}_{Na} m^3 h$$

and

$$\bar{i}_{Na} = \bar{g}_{Na}(E - E_{Na}),$$

where $\bar{g}_{Na}$ is a positive constant whose value remains uncertain [a discussion of upper and lower bounds for $\bar{g}_{Na}$ is given by McAllister et al. (1975, pp. 10–11)] and $E_{Na} = 40$ mV. Also, the activation variable m and the inactivation variable h satisfy the equations

$$\frac{dm}{dt} = (\alpha_m)(1 - m) - (\beta_m)m,$$

$$\frac{dh}{dt} = (\alpha_h)(1 - h) - (\beta_h)h,$$

where the functions $\alpha_m, \beta_m, \alpha_h, \beta_h$ are given in Table 4.4. The inward current i_{si}, which has slower kinetics than i_{Na} and is carried at least partly by calcium ions, is governed by considerations somewhat more complicated than Hodgkin–Huxley formalism. A sizeable fraction of i_{si} is not inactivated even during prolonged depolarization. The existence of residual i_{si} is expressed by McAllister et al. by the following description of i_{si}:

$$i_{si} = \bar{g}_{si}(E - E_{si})df + \bar{g}_{si}^{-1}(E - E_{si})d',$$

where

$$\bar{g}_{si} = 0.8 \text{ m}\Omega^{-1}/\text{cm}^2,$$

$$\bar{g}'_{si} = 0.04 \text{ m}\Omega^{-1}/\text{cm}^2,$$

$$E_{si} = +70 \text{ mV},$$

$$d' = \{1 + \exp[-0.15(E + 40)]\}^{-1},$$

and d and f are activation and inactivation variables, respectively, that satisfy the usual Hodgkin–Huxley equations with $\alpha_d, \beta_d \, \alpha_f, \beta_f$ as given in Table 4.4.

Next, there are three time-dependent outward potassium currents that will be denoted by $i_{K_2}, i_{x_1}, i_{x_2}$. None of them resembles the squid potassium current from a quantitative point of view. However, like the squid potassium current, each of these currents can be described with an activation variable, but without an inactivation variable. The current i_{K_2}, which is called the pacemaker current, is described by

$$i_{K_2} = \bar{i}_{K_2}s,$$

where

$$\bar{i}_{K_2} = \frac{2.8\{\exp[0.04(E + 110)] - 1\}}{\exp[0.08(E + 60)] + \exp[0.04(E + 60)]}$$

Table 4.4.

		Current components that show inactivation					
Component	α	β	$\bar{\alpha}$ (ms^{-1})	$\bar{\beta}$ (ms^{-1})	$\bar{E}$ (mV)	$\bar{g}$ (mΩ^{-1}/cm^2)	E_{rev} (mV)
i_{Na}	$\alpha_m = \dfrac{\bar{\alpha}_m\left(E - \bar{E}_m\right)}{1 - \exp - \dfrac{\left(E - \bar{E}_m\right)}{10}}$	$\beta_m = \bar{\beta}_m \exp - \dfrac{\left(E - \bar{E}_m\right)}{17.86}$	1.0	9.86	-47	150	$+40$
	$\alpha_h = \bar{\alpha}_h \exp - \dfrac{\left(E - \bar{E}_h\right)}{5.43}$	$\beta_h = \bar{\beta}_h\left[1\exp - \dfrac{\left(E - \bar{E}_h\right)}{12.2}\right]^{-1}$	1.13×10^{-7}	2.5	-10		
i_{si}	$\alpha_d = \dfrac{\bar{\alpha}_d\left(E - \bar{E}_d\right)}{1 - \exp - \dfrac{\left(E - \bar{E}_d\right)}{10}}$	$\beta_d = \bar{\beta}_d \exp - \dfrac{\left(E - \bar{E}_d\right)}{11.26}$	0.002	0.02	-40	0.8	$+70$
	$\alpha_f = \bar{\alpha}_f \exp - \dfrac{\left(E - \bar{E}_f\right)}{25}$	$\beta_f = \bar{\beta}_f\left[1 + \exp - \dfrac{\left(E - \bar{E}_f\right)}{11.49}\right]^{-1}$	0.00253	0.02	-26		
i_{qr}	$\alpha_q = \dfrac{\bar{\alpha}_q\left(E - \bar{E}_q\right)}{1 - \exp - \dfrac{E - \bar{E}_q}{10}}$	$\beta_q = \bar{\beta}_q \exp - \dfrac{\left(E - \bar{E}_q\right)}{11.26}$	0.008	0.08	0	2.5	-70
	$\alpha_r = \bar{\alpha}_r \exp - \dfrac{\left(E - \bar{E}_r\right)}{25}$	$\beta_r = \bar{\beta}_r\left[1 + \exp - \dfrac{\left(E - \bar{E}_r\right)}{11.49}\right]^{-1}$	2.08×10^{-5}	0.00	-26		
	$\alpha'_r = \bar{\beta}'_r \exp - \dfrac{\left(E - E'_r\right)}{17}$	$\beta'_r = \bar{\beta}'_r\left[1 + \exp - \dfrac{\left(E - E'_r\right)}{8}\right]^{-1}$	1.93×10^{-4}	0.033	-30		

Source: R. E. McAllister, D. Noble, and R. W. Tsien, Reconstruction of the electrical activity of cardiac Purkinje fibres, *J. Physiol.* 251:7 (1975). Reproduced with permission of Cambridge University Press.

and

$$\frac{ds}{dt} = \alpha_s(1-s) - \beta_s s,$$

where α_s and β_s are given in Table 4.5. The currents i_{x_1} and i_{x_2}, called the plateau currents, are governed by the following equations:

$$i_{x_1} = \bar{i}_{x_1} x_1,$$

$$\bar{i}_{x_1} = \frac{(1.2)(\exp[0.04(E+95)] - 1)}{\exp[0.04(E+45)]},$$

$$\frac{dx_1}{dt} = \alpha_{x_1}(1-x_1) - \beta_{x_1}(x_1),$$

$$i_{x_2} = \bar{i}_{x_2} x_2,$$

where

$$\bar{i}_{x_2} = 25 + 0.385E,$$

and

$$\frac{dx_2}{dt} = \alpha_{x_2}(1-x_2) - (\beta_{x_2})x_2,$$

where $\alpha_{x_1}, \beta_{x_1}, \alpha_{x_2}, \beta_{x_2}$ are given in Table 4.5. There is also a transient outward current i_{qr} carried by chloride ions that is described by

$$i_{qr} = qr(\bar{g}_{qr})(E - E_{Cl}),$$

where

$$\bar{g}_{qr} = 2.5 \text{ m}\Omega^{-1}/\text{cm}^2,$$

$$E_{Cl} = -70 \text{ mV},$$

and q and r are activation and inactivation variables, respectively, that satisfy the equations

$$\frac{dq}{dt} = \alpha_q(1-q) - \beta_q(q),$$

$$\frac{dr}{dt} = \alpha_r(1-r) - \beta_r(r),$$

and $\alpha_q, \beta_q, \alpha_r, \beta_r$ are given in Table 4.4.

Table 4.5.

Component	α	β	$\bar{\alpha}$ (ms^{-1})	$\bar{\beta}$ (ms^{-1})	$\bar{E}$ (mV)	$\bar{g}$ (mΩ^{-1}/cm^2)	E_{rev} (mV)
		Slow outward K currents					
i_{K_2}	$\alpha_8 = \dfrac{\bar{\alpha}_8(E - \bar{E}_8)}{1 - \exp - \dfrac{(E - \bar{E}_8)}{5}}$	$\beta_8 = \bar{\beta}_8 \exp - \dfrac{(E - \bar{E}_8)}{14.93}$	0.001	5.0×10^{-5}	-52	(Rectifying)	-110
i_{x_1}	$\alpha_{x_1} = \dfrac{\bar{\alpha}_{x_1} \exp \dfrac{E + 50}{12.1}}{1 + \exp \dfrac{E + 50}{17.5}}$	$\beta_{x_1} = \dfrac{\bar{\beta}_{x_1} \exp - \dfrac{E + 20}{16.67}}{1 + \exp - \dfrac{E + 20}{25}}$	5×10^{-4}	0.0013	—	(Rectifying)	-95
i_{x_2}	$\alpha_{x_2} = \bar{\alpha}_{x_2} \left[1 + \exp - \dfrac{E + 19}{5}\right]^{-1}$	$\beta_{x_2} = \dfrac{\bar{\beta}_{x_2} \exp - \dfrac{E + 20}{16.67}}{1 + \exp - \dfrac{E + 20}{25}}$	1.27×10^{-4}	3×10^{-4}	—	0.385	-65

Source: R. E. McAllister, D. Noble, and R. W. Tsien, Reconstruction of the electrical activity of cardiac Purkinje fibres, *J. Physiol.* 251:8 (1975). Reproduced with permission of Cambridge University Press.

Finally there are several background currents or leak currents that are time-independent. So far these have not been analyzed very well by experimental means, but McAllister et al. propose the following tentative description. There is an outward background current denoted by i_{K_1} that is carried mainly by potassium ions and described by

$$i_{K_1} = \frac{\bar{i}_{K_2}}{2.8} + (0.2)(E + 30)\{1 - \exp[-0.04(E + 30)]\}^{-1},$$

where $\bar{i}_{K_2}$ is as defined earlier. There is an inward background current $i_{Na,b}$ that is probably carried largely by sodium ions and described by

$$i_{Na,b} = \bar{g}_{Na,b}(E - E_{Na}),$$

where

$$\bar{g}_{Na,b} = 0.105,$$

$$E_{Na} = 40 \text{ mV}.$$

Finally there is a background current $i_{Cl,b}$ that is carried by chloride ions and described by

$$i_{Cl,b} = \bar{g}_{Cl,b}(E - E_{Cl}),$$

where the following, somewhat arbitrary, value is assigned to $g_{Cl,b}$:

$$\bar{g}_{Cl,b} = 0.01 \text{ m}\Omega^{-1}/\text{cm}^2$$

and

$$E_{Cl} = 0.70 \text{ mV}.$$

The equation relating membrane potential E and current i_1 across the membrane is

$$\frac{dE}{dt} = \frac{i_1}{C},$$

where C is the capacitance of the Purkinje fiber and is ~ 12 $\mu\text{F}/\text{cm}^2$. Since

$$i_1 = i_{Na} + i_{si} + i_{K_2} + i_{x_2} + i_{qr} + i_{K_1} + i_{Na,b} + i_{Cl,b},$$

then using the expressions we have given for these currents, we obtain for the MNT model the following system of 10 equations:

$$\frac{dE}{dt} = -\frac{1}{C}\Big\{ \bar{g}_{Na}(E - E_{Na})m^3h + \bar{g}_{si}(E - E_{si})df$$

$$+ \bar{g}'_{si}(E - E_{si})d' + \big\{ \bar{i}_{K_2}(E) \big\} s$$

$$+ \big\{ \bar{i}_{x_1}(E) \big\} x_1 + \big\{ \bar{i}_{x_2}(E) \big\} x_2 + qr(\bar{g}_{qr})(E - E_{Cl})$$

$$+ i_{K_1}(E) + i_{Na,\,b}(E) + i_{Cl,\,b}(E) \Big\},$$

where $\bar{i}_{x_1}, \bar{i}_{x_2}, \bar{i}_{K_2}, i_{K_1}, i_{Na,\,b}, i_{Cl,\,b}$ are, as indicated, all functions of E, and

$$\frac{dy}{dt} = (\alpha_y)(1 - y) - (\beta_y)(y),$$

where $y = m, d, s, x_1, x_2, q, h, f, r$ and the functions $\alpha_y(E), \beta_y(E)$ are given in Tables 4.4 and 4.5.

These equations give a quantitative description of the ionic currents, that is, they describe quantitatively the sequence of events that can be described roughly in words as follows: When the membrane potential is at a suitable level (when depolarization to the sodium threshold has occurred), the inward current i_{Na} starts to increase. The inward current i_{si} increases more slowly and the outward current i_{qr} develops. The i_{Na} decreases and there are quick variations in the membrane potential (termed a spike and a notch) followed by a longer period during which the membrane potential stays fairly level on a so-called plateau. Repolarization or hyperpolarization is triggered by the onset of the current i_{x_1} followed by the onset of the currents i_{x_2} and i_{K_2}. Then the current i_{K_2} decays and this allows the inward background current to cause depolarization again to the sodium threshold. (That is, the decrease in current i_{K_2} permits depolarization to the sodium threshold. It is for this reason that i_{K_2} is called the pacemaker current.)

As has already been emphasized, the MNT model is tentative because it does not have as firm an experimental basis as the Hodgkin–Huxley model. A fair judgment of the model requires a study of the discussion given in McAllister et al. (1975). Nevertheless, the model does summarize a considerable amount of quantitative data with accuracy. Moreover, it is very important for a mathematician who is interested in cell electrophysiology to look closely at the MNT model. In the study of strongly nonlinear ordinary differential equations, the difficulties increase very rapidly as the dimension of the system increases. (Even the transition from two- to three-dimensional systems offers very serious obstacles.) Consequently, mathematicians tend, where possible, to deal with systems of low dimension. Models of dimension two can be useful in some qualitative studies, as has already been indicated in the discussion of the two-dimensional models in Chapter 3. But the MNT model, which has dimension 10, indicates clearly that more realistic models in physiology require working with high-dimensional systems.

4.4.1.4 *Mathematical analysis of models of the cardiac Purkinje fiber*

We have seen that the physiologically significant mathematical problems that need to be studied for the models of the squid axon, the myelinated axon, and the striated muscle fiber are all about the same. In each of these models, we need to study such phenomena as threshold behavior and refractory period. But the physiological functions of the cardiac Purkinje fiber are very different from those of the squid axon, the myelinated axon, and the striated muscle fiber, and as a result, the physiologically significant mathematical problems that need to be studied are very different.

The numerical analysis of the MNT equations carried out in the original paper [McAllister et al. (1975)] suggests the existence of an oscillatory solution. Indeed, it seems reasonable that the regular spontaneous firing of the Purkinje fibers would be described by a periodic solution of the MNT equations. We would expect, moreover, that this periodic solution would be asymptotically stable with a fairly large region of stability. Experimental evidence actually suggests that there are two distinct oscillations [see Hauswirth, Noble, and Tsien (1969)]. We shall return to a discussion of this problem in Chapter 6.

Thus, the first mathematical problem to be studied for the MNT equations is a very classical question: How do we establish the existence of an asymptotically stable periodic solution? Let us assume that this problem has been solved. Next we consider the mathematical description of the primary function of the Purkinje fibers, that is, the transmission of regular electrical impulses that are initiated in the pacemaker region (at the SA node) and that occur with a higher frequency than the frequency of the spontaneous firings of the Purkinje fiber. In order to obtain such a mathematical description, we must add to the MNT equations a term describing the periodic electrical impulses that originate in the pacemaker. If we write the MNT equations in vector form as

$$\dot{x} = P(x) \tag{4.5}$$

and if T is the period of the asymptotically stable periodic solution of the MNT equations, the existence of which is assumed to have been established, then the pacemaker impulses can be described by adding to (4.5) a term of the form $F(t)$, where $F(t)$ has period $T_1 < T$. That is, $F(t)$ is an analytic description of the pacemaker impulse. Thus to obtain a description of the occurrence of regular or periodic impulses, we must search for a solution of

$$\dot{x} = P(x) + F(t), \tag{4.6}$$

which has period T_1. Now this problem is the mathematical formulation of a well-known problem in mechanics and electric circuit theory: the problem of entrainment of frequency. The standard technique for dealing with the problem was introduced by Poincaré and consists of replacing (4.6) with the equation

$$\dot{x} = P(x) + G(t, \varepsilon), \tag{4.7}$$

where ε is a nonnegative parameter, $G(t, \varepsilon)$ has the period $T(\varepsilon)$ as a function of t, and $G(t, 0) \equiv 0$. Under rather general conditions, it can then be proved that for all ε sufficiently small, (4.7) has a nontrivial (i.e., nonconstant) solution of period $T(\varepsilon)$.

In Chapter 5, we will discuss in detail both the problem of showing that (4.5) has an asymptotically stable periodic solution and the problem of entrainment of frequency.

4.4.2   The Beeler–Reuter model of ventricular myocardial fiber

A kind of companion piece to the paper of McAllister et al. (1975) is the model of electrical activity in mammalian ventricular myocardial fibers proposed by Beeler and Reuter (1977).

This model, like the models previously described, is based on data obtained from voltage-clamp experiments.

Following Beeler and Reuter, we let V_m denote the membrane potential. The mathematical model consists of the following set of equations, the first of which is

$$\frac{dV_m}{dt} = -\frac{1}{C_m}(i_{K_1} + i_{x_1} + i_{Na} + i_s - i_{ext}), \tag{4.8}$$

where $C_m = 1$, and

$$i_{K_1} = 0.35 \left\{ \frac{\exp[0.04(V_m + 85)] - 1}{\exp[0.08(V_m + 53)] + \exp[0.04(V_m + 53)]} \right.$$
$$\left. + 0.2\frac{(V_m + 23)}{1 - \exp[-0.04(V_m + 23)]} \right\},$$

$$i_{x_1} = (x_1) \left\{ \frac{0.08(\exp[0.04(V_m + 77)] - 1)}{\exp[0.04(V_m + 35)]} \right\},$$

and

$$i_{Na} = (\bar{g}_{Na}m^3hj + g_{NaC})(V_m - E_{Na}),$$

in which

$$\bar{g}_{Na} = 4 \text{ m}\Omega^{-1}/\text{cm}^2,$$
$$g_{NaC} = 0.003 \text{ m}\Omega^{-1}/\text{cm}^2,$$
$$E_{Na} = 50 \text{ mV}.$$

The current carried mainly by Ca ions, denoted by i_s, is

$$i_s = \bar{g}_s df(V + 82.3 + 13.0287 \ln[\text{Ca}]_i),$$

where $\bar{g}_s = 0.09 \text{ m}\Omega^{-1}/\text{cm}^2$ and $[\text{Ca}]_i$ is in moles and refers to calcium ion concentration in the interior of the fiber. The term i_{ext} is a given function of time that depends on the experiment being described. See Beeler and Reuter (1977, p. 180).

The second equation in the system is

$$\frac{d}{dt}[\text{Ca}]_i = -10^{-7}i_s + 0.07(10^{-7} - [\text{Ca}]_i) \tag{4.9}$$

Table 4.6. C, *defining function and values for rate constants* (α *or* β)
$\alpha = (C_1 \exp[C_2(V_m + C_3)] + C_4(V_m + C_5))/\exp[C_6(V_m + C_3)] + C_7)$.

Rate constant (ms^{-1})	C_1 (ms^{-1})	C_2 (mV^{-1})	C_3 (mV)	C_4 $[(\text{mV ms})^{-1}]$	C_5 (mV)	C_6 (mV^{-1})	C_7
α_{x_1}	0.0005	0.083	50	0	0	0.057	1
β_{x_1}	0.0013	-0.06	20	0	0	-0.04	1
α_m	0	0	47	-1	47	-0.1	-1
β_m	40	-0.056	72	0	0	0	0
α_h	0.126	-0.25	77	0	0	0	0
β_h	1.7	0	22.5	0	0	-0.082	1
α_j	0.055	-0.25	78	0	0	-0.2	1
β_j	0.3	0	32	0	0	-0.1	1
α_d	0.095	-0.01	-5	0	0	-0.072	1
β_d	0.07	-0.017	44	0	0	0.05	1
α_f	0.012	-0.008	28	0	0	0.15	1
β_f	0.0065	-0.02	30	0	0	-0.2	1

Source: G. W. Beeler, and H. Reuter, Reconstruction of the action potential of ventricular myocardial fibers, *J. Physiol.* 268:181 (1977). Reproduced with permission of Cambridge University Press.

and the remaining six equations are

$$\frac{dy}{dt} = \alpha_y(1 - y) - \beta_y(y),$$

where $y = x_1$, m, h, j, d, and f, and the functions α_y and β_y are listed in Table 4.6.

Beeler and Reuter (1977) compare this model with the MNT model and from their discussion, it follows that the types of mathematical problems that arise in the study of this model are very similar to those that arise in the study of the cardiac Purkinje fiber models we have described. Here we merely point out the existence of oscillatory phenomena [Beeler and Reuter (1977, p. 200ff)] and the consequent question of the existence of periodic solutions.

5

Mathematical theory

5.1 Introduction

So far we have described a number of mathematical models of electrically excitable cells and, at least for some of the models, we have indicated the kind of mathematical questions and analysis that arise in work on these models. The next major step is to describe in detail how the mathematical analysis can be carried out. In order to obtain this description, we will first give some theory of differential equations. By doing this, we will introduce the mathematical language that is appropriate for discussing the problems that are of concern to us. We also describe some mathematical techniques and results that will be useful in the study of the models.

5.2 Basic theory

In this section we describe some basic properties of solutions of differential equations. These are used very frequently in the analysis of all the physiological models.

5.2.1 Existence theorems and extension theorems

It is reasonable that our first concern should be for the existence of solutions. To a reader whose experience with differential equations is limited to an introductory course following calculus, such a concern may seem unnecessarily fussy. In a first course in differential equations various techniques for computing solutions are described and it might be expected that our chief concern would be to summarize such computational techniques; but all the physiological models in which we are interested are nonlinear systems, and, consequently, it is usually impossible to get explicit expressions for the solutions (i.e., closed solutions). Frequently we must settle for an approximation to the solution. However, in order to deal with an approximation, we must know that a solution exists. Thus we are seriously concerned with the question of existence, and our first step will be to state a version of the basic existence theorem. Let x denote a real n vector, that is,

$$x = (x_1, \ldots, x_n),$$

where $x_1, \ldots, x_n$ are real numbers and let $|x|$ be defined by

$$|x| = \sum_{j=1}^{n} |x_j|.$$

Similarly, let

$$f(t, x) = (f_1(t, x), \ldots, f_n(t, x)),$$

where each $f_j(t, x)$ (with $j = 1, \ldots, n$) is a continuous real-valued function that is defined on the set

$$R = \{(t, x)/|t - t_0| \le a, |x - x_0| \le b\},$$

where t_0, x_0 are fixed and a and b are fixed positive numbers. Let $M = \max_{(t, x) \in R} |f(t, x)|$ on R and let α be the minimum of b/M and a. We also assume that f satisfies a Lipschitz condition with respect to x in R; that is, we assume that there is a positive number k such that if $(t, x^{(1)})$ and $(t, x^{(2)})$ are points in R, then

$$|f(t, x^{(1)}) - f(t, x^{(2)})| \le k|x^{(1)} - x^{(2)}|.$$

(It is easy to show, by using the mean value theorem, that if the partial derivatives $\partial f_i / \partial x_j$, where $i, j = 1, \ldots, n$, exist and are bounded on R, then f satisfies a Lipschitz condition with respect to x in R.

Existence theorem. *The n-dimensional system*

$$\frac{dx}{dt} = f(t, x) \quad \text{or} \quad \begin{aligned} \frac{dx_1}{dt} &= f_1(t, x_1, \ldots, x_n), \\ \frac{dx_2}{dt} &= f_2(t, x_1, \ldots, x_n), \\ \frac{dx_n}{dt} &= f_n(t, x_1, \ldots, x_n), \end{aligned} \tag{5.1}$$

(with the components written out) has a unique solution $x(t, t_0, x_0)$ such that

$$x(t_0, t_0, x_0) = x_0.$$

The solution $x(t, t_0, x_0)$ is defined for all $t \in [t_0 - \alpha, t_0 + \alpha]$ and

$$|x(t, t_0, x_0) - x_0| \le M\alpha$$

for all $t \in [t_0 - \alpha, t_0 + \alpha]$.

[Following a usual convention we will refer to (5.1) sometimes as an equation and sometimes as a system of equations.]

This theorem, although useful, leaves open a number of questions. Apart from the fact that, as stated, the theorem gives no information about how to compute the solution, there is the question of how large the domain of the solution is. The set R described before the statement of the theorem is not unique. For example, another set R with larger a and b and larger α might exist. Thus, the solution would be extended so that it was defined over a larger interval $[t_0 - \alpha, t_0 + \alpha]$. On the other hand, we cannot be certain of how large a domain can be found. The following simple example shows that the domain may be severely limited. Consider the scalar equation

$$\frac{dx}{dt} = x^2.$$

A solution of this equation that satisfies the condition $t_0 = 0$, $x_0 = 1$ is given by $x(t) = -1/(t - 1)$. As t increases from 0 toward 1, this solution decreases without bound and the solution is not defined at $t = 1$.

Now for convenience, we introduce a precise definition of boundedness of solutions.

Definition. A solution $x(t)$ of (5.1) is *bounded for $t \ge t_0$* if there exists a positive number M such that

$$|x(t)| < M$$

for all $t \ge t_0$ for which $x(t)$ is defined.

It seems reasonable to expect that if a differential equation is a physiological model, then its solutions should be defined for all t greater than some fixed value, that is, for all time later than a certain instant, and that its solutions should remain bounded. (If a solution suggests that membrane potential increases without bound as $t \to 5$ or even as t increases without bound, the validity of the model is indeed questionable!) These two rather natural require-ments on the solutions, that they are defined for all t greater than a fixed value and that they are bounded, are closely connected. It can be proved that if a solution remains bounded for all t past a certain value for which the solution is defined, then the solution is defined for all t past that certain value. In order to state this result precisely, we need one definition. This definition is just a formal description of the property that a solution has been extended as far as possible or has the largest possible domain.

Definition. If $x(t)$ is a solution on (a, b) of $dx/dt = f(t, x)$ and $y(t)$ is a solution on (α, β) of the same equation, and $(\alpha, \beta) \subset (a, b)$ and if, for $t \in (a, b)$, $y(t) = x(t)$, then $y(t)$ is an *extension* of $x(t)$. (*Note:* a is a real number or $-\infty$ and b is a real number or $+\infty$. Similar conditions hold on α and β.)

Definition. If $x(t)$ is a solution on (a, b) of $dx/dt = f(t, x)$ and if $x(t)$ is such that any extension $y(t)$ of $x(t)$, where $y(t)$ is a solution on (α, β) of $dx/dt = f(t, x)$, has the property that $(\alpha, \beta) = (a, b)$, then $x(t)$ is a *maximal solution* of $dx/dt = f(t, x)$.

Extension theorem. *Suppose that $x(t)$ is a maximal solution of*

$$\frac{dx}{dt} = f(t, x),$$

where f is continuous for all real t, all $x \in R^n$, and f satisfies a Lipschitz condition in x in each bounded set E in $R \times R^n$, that is, each set E for which there is a positive number A such that if $(t, x) \in E$, then

$$|t| + |x| < A.$$

Suppose further that (a, b) is the domain of $x(t)$ and that there exists a number $c \in (a, b)$ and a positive number B such that for all $t \in (c, b)$,

$$|x(t)| \le B.$$

Then $b = +\infty$.

Thus the problem of showing that the solutions to be studied have the desirable properties of being defined for all time past a certain instant and of being bounded is reduced to the single problem of showing that the solutions are bounded. As we shall see later in Chapter 6, when mathematical analyses of various physiological models are made, proving boundedness of solutions is often done quite easily. It is often easy to prove that all the physiologically significant solutions enter and remain in a particular bounded set. Then solution $x(t)$ is defined for all $t > a$, that is, $b = +\infty$.

5.2.2 Autonomous systems

An important special case of the differential equation

$$\frac{dx}{dt} = f(t, x)$$

is the equation

$$\frac{dx}{dt} = g(x),$$

that is, an equation in which the right-hand side is not a function of the independent variable t. Such an equation is called an autonomous equation. Autonomous equations occur often in mechanics and electrical circuit theory; all the physiological models we have considered so far are autonomous equations. A crucially important property of autonomous equations is that their solutions can be studied geometrically in a particularly nice way. In order to describe this property informally, we consider the two-dimensional autonomous system

$$\frac{dx}{dt} = f(x, y),$$

$$\frac{dy}{dt} = g(x, y).$$

A solution of this system is a pair of differentiable functions $(x(t), y(t))$. The pair of functions $x(t)$ and $y(t)$ describe a curve in the xy plane. It is easy to show [see, e.g., Cronin (1980b)] that if c is a constant, then $(x(t + c), y(t + c))$ is also a solution of the system that describes the same curve. Also $(x(t + c), y(t + c))$ is the only kind of solution that describes that same curve. [Here we are using the term "curve" in the intuitive sense, that is, to refer to the point set or geometric configuration. If we were to use the term "curve" in the rigorous mathematical sense, then we would say that $(x(t+c), y(t+c))$ is an element of the equivalence class that constitutes the curve.] From the uniqueness condition in the conclusion of the basic existence theorem, it follows that no solution curve crosses itself and that the intersection of two distinct solution curves is the null set. (It is easy to show with examples that these results do not hold if the system is not autonomous.) As a result, it is often enlightening to study the curves instead of the solutions themselves. The xy plane is often called the phase plane and the graphs of the curves are termed the phase plane portrait. Later we will see how important this phase plane technique and its n-dimensional extension are for the study of physiological models.

5.2.3   Equilibrium points
5.2.3.1   *Definition of equilibrium point*
The simplest class of solutions of a system of ordinary differential equations is the constant solutions. If we consider an n-dimensional autonomous system

$$\frac{dx_1}{dt} = f_1(x_1, \ldots, x_n),$$

$$\frac{dx_2}{dt} = f_2(x_1, \ldots, x_n),$$

$$\vdots$$

$$\frac{dx_n}{dt} = f_n(x_1, \ldots, x_n),$$

(5.2)

and if $\bar{x}_1, \ldots, \bar{x}_n$ are real constants such that for $j = 1, \ldots, n$,

$$f_j(\bar{x}_1, \ldots, \bar{x}_n) = 0,$$

then $(x_1(t), \ldots, x_n(t))$, where $x_j(t)$ is defined by

$$x_j(t) = \bar{x}_j \quad \text{for all } t,$$

is a solution of the system. It is, of course, a very special solution, each of whose components is a constant. It is called a critical point, a singular point, or an equilibrium point. Since the terms "critical" and "singular" have other meanings in mathematics, the term equilibrium point will be used here.

(We have defined the notion of equilibrium point only for autonomous systems. The definition can be extended to include nonautonomous systems: A point $\bar{x}$ is an equilibrium point of the system $dx/dt = f(t, x)$ if $f(t, x) = 0$ for all t. However, in our work, we will be concerned only with equilibrium points of autonomous systems.)

5.2.3.2 *The two-dimensional case*

Although equilibrium points seem like highly specialized solutions, their locations and the behavior of solutions near them give very useful information about the general behavior of solutions of the system. Our next step is to summarize briefly some well-known results concerning equilibrium points. We begin with the two-dimensional case for which fairly complete results can be obtained.

Linear homogeneous systems. Consider first the linear homogeneous system

$$\frac{dx}{dt} = ax + by,$$
$$\frac{dy}{dt} = cx + dy,$$

(5.3)

where a, b, c, d are real constants and it is assumed that

$$\det\begin{bmatrix} a & b \\ c & d \end{bmatrix} = ad - bc \neq 0.$$

Since this determinant is nonzero, then $(0, 0)$ is the only equilibrium point. We assume further that the matrix

$$M = \begin{bmatrix} a & b \\ c & d \end{bmatrix}$$

is in canonical form. Then analysis of an elementary nature [see, e.g., Cronin (1980b)] yields easily the phase plane portraits. Let λ_1, λ_2 be the eigenvalues of the matrix M. Then the phase portraits depend on the relationship between λ_1 and λ_2. Sketches of the phase portraits and their names are as follows:

Case I. λ_1, λ_2 are real, unequal, and have the same sign.

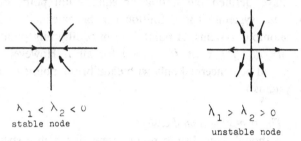

$$\lambda_1 < \lambda_2 < 0$$
stable node

$$\lambda_1 > \lambda_2 > 0$$
unstable node

Case II. λ_1, λ_2 are real and equal and $M = \begin{bmatrix} \lambda & 0 \\ 0 & \lambda \end{bmatrix}$.

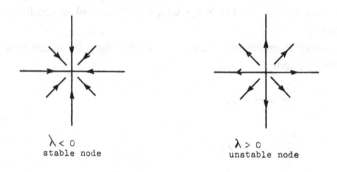

$$\lambda < 0$$
stable node

$$\lambda > 0$$
unstable node

Case III. λ_1, λ_2 are real and equal and $M = \begin{bmatrix} \lambda & 1 \\ 0 & \lambda \end{bmatrix}$.

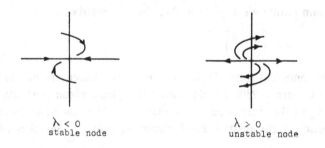

$$\lambda < 0$$
stable node

$$\lambda > 0$$
unstable node

Case IV. $\lambda_2 < 0 < \lambda_1$.

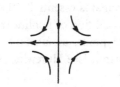

saddle point

Case V. λ_1, λ_2 are complex conjugate numbers, that is, $\lambda_1 = \alpha + i\beta$, $\lambda_2 = \alpha - i\beta$.

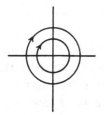
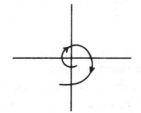

$\lambda_1 = i\beta,\ \beta > 0$

center or vortex

$\lambda_1 = \alpha + i\beta,\ \alpha > 0,\ \beta > 0$
unstable spiral point or
unstable focus

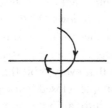

$\lambda_1 = \alpha + i\beta,\ \alpha < 0,\ \beta > 0$

stable spiral point or
stable focus

If matrix M is not in canonical form, then further elementary analysis shows that the phase portraits are the same as those just described except that they are distorted by a transformation that is a nonsingular linear transformation.

Nonlinear systems. It can be shown that if nonlinear, higher-order terms are added to the right-hand side of (5.3), then in a neighborhood of the origin, the phase portrait remains essentially the same except in the case of a center where the addition of nonlinear terms may result in a center, a stable spiral point, an unstable spiral point, or a more complicated phase portrait [see Hurewicz (1958)].

Finally, we consider the general case

$$\frac{dx}{dt} = P(x, y),$$

$$\frac{dy}{dt} = Q(x, y),$$

(5.4)

where $P(x, y)$ and $Q(X, Y)$ are power series in x and y and $(0,0)$ is an isolated equilibrium point, that is, there is a neighborhood of $(0,0)$ such that the only equilibrium point in the neighborhood is $(0,0)$. We have just seen what the phase portrait looks like if the coefficient matrix of the linear terms in $P(x, y)$ and $Q(x, y)$ is nonsingular. But so far we have no information about the behavior of the orbits if P or Q is a series that starts with terms of degree higher than 1. There is a remarkably simple answer to this question, which can be described roughly as follows. If $(0,0)$ is an isolated equilibrium point of (5.4), then either $(0,0)$ is a spiral [i.e., the orbits spiral toward $(0,0)$ as t increases (decreases)] or a center [there is a neighborhood N of $(0,0)$ such that every orbit that has a nonempty intersection with N is a simple closed curve] or a neighborhood of $(0,0)$ can be divided into a finite number of "sectors" of the following types:

Type I. Fan or parabolic sector.

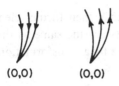

(0,0)   (0,0)

Type II. Hyperbolic sector.

Type III. Elliptic sector.

In each of these sketches, the arrows indicate the direction of increasing time. Note that the only essentially new configuration that arises when nonlinear equilibrium points are studied is the elliptic sector.

For a precise statement of this result and a proof, see Lefschetz (1963, Chap. X). For a similar discussion see Nemytskii and Stepanov (1960, Chap. II).

5.2.3.3 *The index of an equilibrium point*

A beautiful and sometimes useful concept that is introduced in the study of equilibrium points is the *index of the equilibrium point*. We describe first an intuitive version of the definition of index for equilibrium points in the two-dimensional case.

Definition. Let $P(x, y), Q(x, y)$ be real-valued continuous functions on an open set D in the xy plane. Let $V(x, y)$ be the vector field

$$V(x, y) = (P(x, y), Q(x, y)),$$

that is, V is a function with domain D and range contained in R^2. If $(x, y) \in D$, then $V(x, y)$ is the point or vector $(P(x, y), Q(x, y))$.

An *equilibrium point* of $V(x, y)$ is a point $(x_0, y_0) \in D$ such that

$$P(x_0, y_0) = Q(x_0, y_0) = 0.$$

Let C be a simple closed curve in D such that no point of C is an equilibrium point of V, and consider how the vector associated with a point $(x_1, y_1) \in C$ changes as (x_1, y_1) is moved counterclockwise around C. The vector is rotated through an angle $j(2\pi)$, where j is an integer, positive or negative or zero. The number j is the *index of C with respect to V(x, y)*.

It is easy to see that this "definition" is not precisely formulated because several of the terms used have only an intuitive meaning. For example, what does it mean to "move counterclockwise" on C? As long as C is an easily visualized curve like a circle or ellipse, it is easy to describe precisely what we mean by moving counterclockwise. But if C is a sufficiently "messy" curve, then it is not at all clear which is the counterclockwise direction on the curve. Second, although for easily visualized curves C it is intuitively clear how to measure the angle through which the vector is rotated, no general and precise method for describing how to measure the angle has been given. Thus our "definition" is far from rigorous. An efficient way to give a rigorous definition is to formulate the definition in terms of topological degree, a procedure that we will shortly describe.

Having obtained the definition of the index of C with respect to $V(x, y)$, one can then prove that if C is continuously moved or deformed in such a way that it does not cross any equilibrium point of V during the continuous deformation, then the index j remains constant during the deformation.

Now suppose that (x_0, y_0) is an isolated equilibrium point of V, that is, suppose that there exists a circular neighborhood N of (x_0, y_0) such that there are no equilibrium points of V in N except (x_0, y_0). Let C_1, C_2 be simple closed curves in N such that (x_0, y_0) is in the interior of C_1 and is the interior of C_2. It is not difficult to show that C_1 can be continuously deformed into C_2 without crossing any equilibrium point of V. Hence the index of C_1 equals the index of C_2, and indeed the index j is independent of the simple closed curve C provided C is contained in N and (x_0, y_0) is in the interior of C. Hence we may formulate the following definition.

Definition. The index j is the index of the isolated equilibrium point (x_0, y_0).

We have already seen that this definition is not based on rigorous considerations. Also it is desirable to have a definition that is valid in the n-dimensional case. Both of these difficulties can be dealt with by defining the index of an equilibrium point in terms of topological degree or Brouwer degree. Next we describe very briefly how this is done. Our description is far from complete because no attempt is made to define Brouwer degree. [For a description of the Brouwer degree, see Krasnosel'skii (1964), Lloyd (1978), or Cronin (1980b).]

Definition. If p is an isolated equilibrium point, the *index of p* is

$$\deg_B[f, B^n, 0],$$

where $\deg_B$ is the Brouwer degree, and B^n is a ball with center p such that $B^n \subset N$.

If $n = 2$, this definition agrees with the informally described definition previously given. For arbitrary n, if the matrix

$$\left[\frac{\partial f_i}{\partial x_j}(p)\right]$$

is nonsingular, it follows that the index of p is

$$\text{sign det}\left[\frac{\partial f_i}{\partial x_j}(p)\right].$$

By using this topological version of the definition of index of an equilibrium point, one can obtain the following existence theorem, which is useful in the study of physiological models.

Theorem 1. *Given the autonomous system*

$$\frac{dx_1}{dt} = f_1(x_1, \dots, x_n)$$

$$\vdots \qquad\qquad\qquad (5.5)$$

$$\frac{dx_n}{dt} = f_n(x_1, \dots, x_n),$$

where f_j has continuous first derivatives in R^n, suppose that $\overline{U}$ is the closure of an open ball $U = \{ P /\lvert p - p_0 \rvert < r \}$ with center p_0 and radius r, and that U is such that if a solution of system (5.5) passes through a point in U, then for all later time t, the solution stays in $\overline{U}$. Suppose also that there are no equilibrium points in $\overline{U} - U$. Then (5.5) has at least one equilibrium point in U.

Proof. See Cronin (1980b, Appendix: Corollary 6.1).

Theorem 2. If, besides the hypotheses in Theorem 1, it is assumed that the indices of the equilibrium points in U all have the same sign, then (5.5) has exactly one equilibrium point in $\overline{U}$.

Proof. See Cronin (1980b, Appendix: Property 3).

5.2.4   Stability and asymptotic stability of solutions
5.2.4.1   *Introduction*
In studying ordinary differential equations that describe biological systems, it is important to determine whether the solutions are stable. Roughly speaking, a solution $x(t)$ is stable if, whenever another solution gets close to $x(t)$, this other solution then stays close to $x(t)$ for all later time. A solution $x(t)$ is asymptotically stable if, whenever another solution gets close to $x(t)$, this other solution then gets closer and closer to $x(t)$ as time increases.

Such a solution can be expected to predict actual behavior of the biological system. On the other hand, a solution that is not stable would probably not give a good prediction of the behavior of the biological system. The reason for this is the presence of small disturbances that are not described or taken into account in the differential equations. It is highly likely that such small disturbances frequently occur in physiological systems, and consequently, if the physiological system is described by a particular solution of the differential equation, we may think of the system as "moving along" the solution until a small disturbance occurs, which "kicks" the system onto a nearby solution. Now if the original solution is stable in the sense that nearby solutions approach it, then the system will "move back" toward the original solution. Thus the original solution will yield a reasonably good prediction of the behavior of the system. However, if the original solution is not stable and if the system is kicked onto a nearby solution by a small disturbance, the nearby solution may move away from the original solution and, as a result, the system will move away from the original solution. Consequently the original solution will give a poor prediction of the behavior of the system.

The notion of stability, which is intuitively very reasonable, requires considerable care if we are to formulate a rigorous definition. Moreover, it is usually quite difficult to determine if a solution is stable and an entire mathematical theory of stability has been developed to deal with this question. The theory is not complete and, from the point of view of applications, the theory is not entirely satisfactory. For the present, we shall simply state formal definitions of stability and indicate very briefly techniques that have been developed to determine whether a given solution is stable.

5.2.4.2 *Asymptotic stability*

We restrict ourselves to considering just one stability concept, that is, asymptotic stability, which seems to be particularly appropriate for applications to differential equations that describe biological systems. [For a detailed discussion of the significance of stability of solutions of differential equations, see Cronin (1980b).]

Although it is logically not necessary to give a separate definition of asymptotic stability for the special case of a solution that is an equilibrium point, we give first such a separate definition. There are two reasons for doing so: first, the asymptotic stability of equilibrium points is, in practice, a very important special case; second, the definition of asymptotic stability for an equilibrium point is intuitively clear and serves as a guide to understanding the definition of asymptotic stability for an arbitrary solution.

We consider an autonomous system

$$\frac{dx_1}{dt} = f_1(x_1, \dots, x_n)$$

$$\vdots \tag{5.6}$$

$$\frac{dx_n}{dt} = f_n(x_1, \dots, x_n),$$

where f_j has continuous first derivatives in R^n ($j = 1, \dots, n$).

Definition. An equilibrium point $p = (\bar{x}_1, \ldots, \bar{x}_n)$ of system (5.6) is *asymptotically stable* if: given $\varepsilon > 0$, then there exists $\delta > 0$ such that if $x(t)$ is a solution of (5.6) and $|x(t_0) - p| < \delta$, then for all $t > t_0$, solution $x(t)$ is defined and $|x(t) - p| < \varepsilon$ and

$$\lim_{t \to \infty} x(t) = p.$$

In words, the equilibrium point is asymptotically stable if each solution that gets close enough to p approaches p as a limit as $t \to \infty$ (see the accompanying sketch).

This basic definition of asymptotic stability of an equilibrium point illustrates the fact that the mathematical theory of stability is not entirely realistic. If an equilibrium point is asymptotically stable and if for some values of ε the corresponding $\delta = \delta(\varepsilon)$ has a very small value, say $\delta = (10^{-30})(\varepsilon)$, then for practical purposes, the equilibrium point is unstable. On the other hand, suppose we consider an equilibrium point of a two-dimensional system for which the orbits near the equilibrium point behave as sketched:

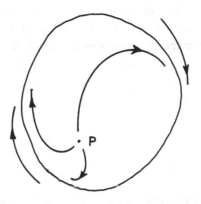

That is, there is a periodic solution (represented by the closed curve) such that all solutions except the equilibrium point p approach the periodic solution. Then the equilibrium point is certainly not asymptotically stable. On the other hand, if the closed curve that represents the periodic solution is contained in a small neighborhood of the equilibrium point, then for practical purposes, the equilibrium point may be regarded as stable.

A fundamental question in stability theory is to obtain criteria for determining if a solution is stable or asymptotically stable. The following theorem is a classical result that is often applied.

Theorem 3. *If the real parts of the eigenvalues of the matrix*

$$M = \left[\frac{\partial f_i}{\partial x_j}(p) \right]$$

are all negative, then the equilibrium point p is asymptotically stable. If M has an eigenvalue with positive real part, then p is not asymptotically stable.

Proof. See LaSalle and Lefschetz (1961) or Cronin (1980).

If $n > 2$, the question of determining whether the eigenvalues of the matrix M have negative real parts is itself a serious problem especially as n gets larger. One important method for solving this problem is the Routh–Hurwitz criterion. We will not describe this criterion in detail. We merely point out that it is described in most texts and that a detailed account including generalizations may be found in Marden (1966).

Theorem 3 leaves unresolved the question of whether p is asymptotically stable if the matrix M is nonsingular and all its eigenvalues either have negative real parts or are pure imaginary. These cases are not just "degenerate cases." They often occur in applications and their study requires a more delicate investigation of the higher-order terms. Sometimes this is done by straightforward (but usually rather laborious) computation. For some classes of equations, the use of Lyapunov functions is an effective method, which largely avoids computation. A lucid and beautiful introduction to Lyapunov functions is given by LaSalle and Lefschetz

(1961). A complete technical account may be found in Cesari (1971).

Now we turn to the definition of asymptotic stability for an arbitrary solution. We consider an n-dimensional system

$$\frac{dx_1}{dt} = f_1(t, x_1, \ldots, x_n),$$

$$\frac{dx_n}{dt} = f_n(t, x_1, \ldots, x_n),$$

$$(5.7)$$

where f_j has continuous first derivatives in $R \times R^n$ ($j = 1, \ldots, n$).

In the following definition $x(t, \bar{t}, \bar{x})$ denotes the solution of (5.7) such that $x(\bar{t}, \bar{t}, \bar{x}) = \bar{x}$.

Definition. Solution $x(t)$ of (5.7) is *stable* if there is a value t_0 such that the following conditions hold.

1. There is a number $b > 0$ such that if

$$|x^1 - x(t_0)| < b,$$

then $x(t, t_0, x^1)$ is defined for all $t \geq t_0$.

2. Given $\varepsilon > 0$, then there exists $\delta \in (0, b)$ such that if

$$|x^1 - x(t_0)| < \delta,$$

then for all $t \geq t_0$,

$$|x(t, t_0, x^1) - x(t)| < \varepsilon.$$

The solution $x(t)$ is *asymptotically stable* if conditions 1 and 2 hold and if also the following condition holds:

3. There exists $\bar{\delta} \in (0, b]$ such that if

$$|x^1 - x(t_0)| < \bar{\delta},$$

then

$$\lim_{t \to \infty} |x(t, t_0, x^1) - x(t, t_0, x^0)| = 0.$$

From the mean value theorem, we have

$$f(t, x + u) - f(t, x) = \left[\frac{\partial f_i}{\partial x_j}(t, x) \right] u + R(t, x, u),$$

where R is a remainder term such that

$$\lim_{|u| \to 0} |R(t, x, u)|/|u| = 0,$$

and a straightforward calculation yields:

Theorem 4. *Solution $x(t)$ is asymptotically stable if and only if 0 is an asymptotically stable equilibrium point of*

$$\frac{du}{dt} = \left[\frac{\partial f_i}{\partial x_j}(t, x(t))\right] u + R[t, x(t), u].$$

Application of the Floquet theory [see Cesari (1971) or Cronin (1980b)] then yields:

Theorem 5. *If f has period T as a function of t, if $x(t)$ is a solution of (5.7), and if $x(t)$ has period T, let $\lambda_1, \ldots, \lambda_n$ be the characteristic multipliers and $\rho_1, \ldots, \rho_m$ the corresponding characteristic exponents of the system*

$$\frac{du}{dt} = \left[\frac{\partial f_i}{\partial x_j}[t, x(t)]\right] u.$$

If $|\lambda_i| < 1$ [or equivalently, $R(\rho_i) < 0$], for $i = 1, \ldots, n$, then $x(t)$ is asymptotically stable. If there exists j such that $|\lambda_j| > 1$ [equivalently $R(\rho_j) > 0$], then $x(t)$ is not stable.

Unfortunately, if $n > 2$, this theorem is more elegant than practical because the problem of computing the characteristic multipliers becomes very difficult unless f and the periodic solution $x(t)$ are given very explicitly.

The following theorem lacks the elegance of the preceding theorem but in some cases, it may be more practical.

Theorem 6. *Suppose that the equation*

$$\frac{dx}{dt} = f(t, x) \tag{5.8}$$

satisfies the following conditions:

1. *Each component of $f(t, x)$ has continuous third derivatives in t and $x_1, \ldots, x_n$, the components of x, at each point of R^{n+1}.*
2. *There exists a bounded open set $U \subset R^n$ such that each solution of (5.8) that intersects ∂U, the boundary of U, is*

"headed into" the interior of U, that is, if $x(t)$ is a solution of (5.8) and $x(t_0) \in \partial U$, then there is a $\delta > 0$ such that if

$$t_0 < t < t_0 + \delta,$$

then $x(t) \in U$, and if

$$t_0 - \delta < t < t_0,$$

then $x(t) \in R^n - \overline{U}$. [This condition implies that no solution of (5.8) escapes U. Also since U is bounded, it follows that if $x(t_0) \in U$, then solution $x(t)$ is defined for all $t \geq t_0$.]

3. Let $P_{t,y}$ denote a matrix such that $|P_{t,y}| = 1$ and

$$P_{t,y} f_x(t, y) P_{t,y}^{-1} = J,$$

where J is the real canonical form of $f_x(t, y)$. [For a description of the real canonical form, see Cronin (1980).] There exists $M > 0$ such that for all $y \in \overline{U}$,

$$\underset{t \in R}{\mathrm{lub}} \left| P_{t,y}^{-1} \right| < M.$$

4. For each $t \in R$, $y \in U$, the eigenvalues of the matrix $f_x(t, y)$ satisfy the following hypotheses:

(i) There exists $r > 0$ such that all of the eigenvalues of $f_x(t, y)$ have real parts that are less than or equal to $-r$. If there exist $y \in \overline{U}$ and $t \geq t_0$ such that one of the eigenvalues of $f_x(t, y)$ has a nonsimply elementary divisor then $r > M$, the constant in condition 3.

(ii) There exist positive numbers $\overline{b}, \overline{c}$ such that

$$r - M - \overline{c} > 0 \quad \text{and} \quad \overline{b} \leq \frac{r - M - \overline{c}}{M}$$

and such that for all $y \in \overline{U}$ and $t \geq t$, the imaginary part of each eigenvalue of $f_x(t, y)$ has absolute value less than $\overline{b}$. In the special case that the eigenvalues of the matrices $f_x(t, y)$ (where $y \in \overline{U}$ and $t \geq t_0$) all have simple elementary divisors, we require only that there exist positive numbers $\overline{b}$ and $\overline{c}$ such that

$$\overline{r} - c > 0 \quad \text{and} \quad \overline{b} \leq \frac{r - \overline{c}}{m}$$

and such that for all $y \in \overline{U}$ and $t \geq t_0$, the imaginary part of each eigenvalue of $f_x(t, y)$ has absolute value less than $\bar{b}$. [*Computable criteria for these hypotheses are given by Marden (1966).*]

(*iii*) *Let* $-q_1, \ldots, -q_n$ *denote the real part of the eigenvalues of* $f_x(t, y)$. *Then for* $j = 2, \ldots, n$, $|q_1 - q_j| < \bar{c}/8$.

Conclusion: *If* $x(t)$ *is a solution of* (5.8) *such that* $x(t_0) \in U$ *for some* t_0, *then* $x(t)$ *is asymptotically stable.*

Proof. See Cronin (1980a).

5.2.4.3 *Phase asymptotic stability*

It is natural to assume that in studying autonomous equations we should use the same definition of asymptotic stability that was used in the discussion of nonautonomous equations. However, it turns out that this is not possible for the most important case, periodic solutions. That is, we have the following theorem [for the proof, see Cronin (1980b)]:

Theorem 7. *If* $x(t)$ *is a nontrivial periodic solution* (*i.e., a periodic solution that is not an equilibrium point*) *of the autonomous equation*

$$\frac{dx}{dt} = f(x),$$

then $x(t)$ *is not asymptotically stable.*

It is rather reasonable to raise, at this point, the question of using Theorem 5 to determine if a periodic solution is asymptotically stable. Actually there is no possibility of applying Theorem 5 because if $\bar{x}(t)$ is a periodic solution of

$$\frac{dx}{dt} = f(x),$$

then

$$\frac{d}{dt}\left[\frac{d\bar{x}}{dt}\right] = \{f_x[\bar{x}(t)]\}\left[\frac{d\bar{x}}{dt}\right].$$

Thus, $d\bar{x}/dt$ is a periodic solution of the equation

$$\frac{du}{dt} = \{f_x[\bar{x}(t)]\}u.$$

It follows that $f_x[\bar{x}(t)]$ has at least one characteristic multiplier equal to 1 [see Cronin (1980b)] and hence the hypothesis of Theorem 5 is violated.

The stability definitions that are frequently used in the study of solutions of autonomous equations are definitions of orbital stability. As the name suggests, these concepts arose in the study of celestial mechanics, and they are not entirely satisfactory for the models that are studied here. Because the choice of a stability definition is important, it is worthwhile to explain this point in detail and we begin by stating the definitions.

Definition. Let $x(t)$ be a solution of the n-dimensional equation

$$\frac{dx}{dt} = f(x), \tag{5.9}$$

where f is defined and has continuous derivatives in R^n such that $x(t)$ has period T. Let C be the orbit of $x(t)$, that is,

$$C = \{x(t)/t \in [0,T]\}$$

or C is the underlying point set of solution $x(t)$. If $p \in R^n$, let

$$d(p,C) = \inf_{q \in C} |p - q|.$$

Solution $x(t)$ is *orbitally stable* iff: given $\varepsilon > 0$, then there exists $\delta > 0$ such that for any solution $x^{(1)}(t)$ of (5.9) for which there is a t_0 such that

$$d\left(x^{(1)}(t_0), C\right) < \delta,$$

it is true that $x^{(1)}(t)$ is defined for all $t > t_0$ and

$$d\left(x^{(1)}(t), C\right) < \varepsilon$$

for all $t \geq t_0$. Solution $x(t)$ is *asymptotically orbitally stable* if $x(t)$ is orbitally stable and there is an $\varepsilon_0 > 0$ such that for any solution $x^{(1)}(t)$ for which there exists t_0 with

$$d\left(x^{(1)}(t_0), C\right) < \varepsilon_0,$$

it is true that

$$\lim_{t \to \infty} d\left(x^{(1)}(t), C\right) = 0.$$

Orbital stability is a meaningful property of a solution only if the orbit is a simple configuration. In practice only the case in which the orbit is a simple closed curve (so that the solution is periodic) is considered. That is why in the definition of orbital stability, we imposed the hypothesis that $x(t)$ has period T. Orbital stability says roughly that if a solution gets close to the orbit of the given solution, it stays close. However, if $y(t_0)$ is close to C, it does not follow that

$$|y(t) - x(t)| \qquad (5.10)$$

remains small. If the solutions move at different speed, the expression (5.10) may not stay small. In other words, a solution may be orbitally stable but not stable in the sense of the definition of stability in the preceding section on nonautonomous equations. [For an example of a periodic solution that is orbitally stable but not stable, see Hahn (1967, p. 172).]

For our purposes where we would expect the solutions near the "stable" periodic solution to have some kind of approximate period, orbital stability is not enough. Roughly speaking, we want the nearby solutions to have about the same speed as the "stable solution." Now we describe an example that illustrates this desirable behavior and after that we will introduce a definition suggested by this example.

Example.

$$\frac{dx}{dt} = y + x(1 - x^2 - y^2),$$

$$\frac{dy}{dt} = -x + y(1 - x^2 - y^2).$$

In polar coordinates, this system becomes

$$r\frac{dr}{dt} = x\frac{dx}{dt} + y\frac{dy}{dt} = (x^2 + y^2)(1 - x^2 - y^2) = r^2(1 - r^2)$$

or

$$\frac{dr}{dt} = r(1 - r^2) \qquad (5.11)$$

and

$$\frac{d\theta}{dt} = -1. \qquad (5.12)$$

Elementary computations show that the general solution of the system (5.11) and (5.12) is

$$\theta = -t + C,$$

$$r = \frac{1}{[1 + ke^{-2t}]^{1/2}},$$

where C and k are arbitrary real constants. If k is negative, then r is real only if t is large enough so that

$$-1 < ke^{-2t}.$$

If $C = 0$, then in Cartesian coordinates, each solution is $(x_k(t), y_k(t))$, where

$$x_k(t) = \frac{\cos t}{[1 + ke^{-2t}]^{1/2}}, \qquad y_k(t) = \frac{-\sin t}{[1 + ke^{-2t}]^{1/2}}.$$

Note that if $k = 0$, then

$$x_0(t) = \cos t, \qquad y_0(t) = -\sin t.$$

Now, if $k_0 \neq 0$,

$$\left| x_{k_0}(t) - x_0(t) \right| + \left| y_{k_0}(t) - y_0(t) \right|$$

$$< \left\{ 1 - \frac{1}{[1 + k_0 e^{-2t}]^{1/2}} \right\} |\cos t| + |\sin t|$$

and, thus,

$$\lim_{t \to \infty} \left\{ \left| x_k(t) - x_0(t) \right| + \left| y_k(t) - y_0(t) \right| \right\} = 0.$$

But since the system of differential equations is autonomous, then if τ is a nonzero constant, the pair $(\bar{x}_0(t), \bar{y}_0(t))$, where

$$\bar{x}_0(t) = x_0(t + \tau), \qquad \bar{y}_0(t) = y_0(t + \tau),$$

is also a solution and for most values of τ,

$$\lim_{t \to \infty} \left\{ \left| x_k(t) - \bar{x}_0(t) \right| + \left| y_k(t) - \bar{y}_0(t) \right| \right\} \neq 0.$$

Thus, if the phase of solution $(x_0(t), y_0(t))$ is changed, the asymptotic stability property is lost.

This example suggests the following definition, which will be useful.

Definition. Let $x(t)$ be a solution of the n-dimensional equation

$$\frac{dx}{dt} = f(x),\tag{5.13}$$

where f is continuous on R^n. Then $x(t)$ is *uniformly stable* if there exists a constant k such that, given $\varepsilon > 0$, then there exists $\delta > 0$ such that if $u(t)$ is a solution of (5.13) and if there exist t_1, t_2 with $t_2 \geq K$ and such that

$$|u(t_1) - x(t_2)| < \delta.$$

then for all $t \geq 0$,

$$|u(t_1 + t) - x(t_2 + t)| < \varepsilon.\tag{5.14}$$

The solution $x(t)$ is *phase asymptotically stable* if (5.14) is satisfied and also there exists t_3 such that

$$\lim_{t \to \infty} |u(t) - x(t_3 + t)| = 0.$$

[In other words, if $u(t)$ gets close enough to the orbit of $x(t)$ and if the phase of $x(t)$ is suitably chosen, then the distance between the solutions goes to 0 as t increases without bound.]

The concept of phase asymptotic stability has been known for a long time in differential equations, and, as the preceding discussion indicates, it seems particularly well adapted for use in problems in physiology and biology. Phase asymptotically stable solutions have a very important property, which is described in a theorem due to Sell (1966) [see also Cronin (1980b)].

To state Sell's theorem, we need a few definitions.

Suppose that we consider the n-dimensional system

$$\frac{dx}{dt} = f(x).\tag{5.15}$$

Definition. The point $\bar{x}$ is an *ω-limit point* of a solution $x(t)$ of (5.15) if there exists a monotonic increasing sequence $\{t_n\}$ such that

$$\lim_{n \to \infty} t_n = \infty \quad \text{and} \quad \lim_{t_n \to \infty} x(t_n) = x.$$

Definition. The set of all ω-limit points of $x(t)$ is the *Ω-limit set* of $x(t)$ and is denoted by $\Omega[x(t)]$.

[Notice that if $x(t)$ is bounded, then $\Omega[x(t)]$ is nonempty.]

Notation. Let $O[x(t)]$ denote the orbit or underlying point set of the solution $x(t)$, that is,

$$O[x(t)] = \{x(t)/t \text{ in the domain of } x(t)\}.$$

5.2.4.4 *Sell's theorem*

Sell's theorem. *If $x(t)$ is a bounded phase asymptotically stable solution of* (5.13), *then there exists a phase asymptotically stable periodic solution $y(t)$ of* (5.13) *such that*

$$\Omega[x(t)] = O[y(t)].$$

Roughly speaking, Sell's theorem says that a bounded phase asymptotically stable solution approaches a phase asymptotically stable periodic solution. (This periodic solution may be just an equilibrium point.) In view of the earlier discussion, Sell's theorem suggests that the physiologically significant solutions all approach equilibrium points or nontrivial periodic solutions. (We use the term "suggests" here because Sell's theorem is based on concepts from the mathematical theory of stability, and that theory is not entirely realistic.)

5.2.4.5 *Existence of phase asymptotically stable solutions*

Although Sell's theorem provides a large clarification of the picture of how phase asymptotically solutions behave, it raises a serious problem, namely, how to determine whether there exists a phase asymptotically stable solution, especially when we are dealing with a system in which we have very little information about the explicit form of the solutions. Of course if the solution is an equilibrium point, phase asymptotic stability coincides with asymptotic stability and the familiar criterion for asymptotic stability (Theorem 3) is applicable. But if we want to use Sell's theorem to search for nontrivial periodic solutions, then we must establish the existence of nontrivial phase asymptotically stable solutions. In order to do this, we use the same kind of technique that is described in Theorem 6. That is, the eigenvalues of the matrix

$$\left[\frac{\partial f_i}{\partial x_j}(x)\right]$$

are examined at all points x that are contained in an orbit or, more practically, at all points x in a set from which a solution cannot escape. Considerations of this kind yield the following theorem.

Theorem 8. *Let*

$$\frac{dx}{dt} = f(x) \tag{5.16}$$

be an n-dimensional autonomous system such that f has continuous second derivatives at each point of R^n. Suppose there exists a bounded open set $U \subset R^n$ such that the following conditions hold:

1. *Each solution of (5.16) that intersects ∂U is headed into the interior of U, that is, if $x(t)$ is a solution of (5.16) and $x(t_0) \in \partial U$, then for all $t \geq t_0$,*

$$x(t) \in U.$$

2. *There is no equilibrium point of (5.16) in U.*
3. *For all $x \in U$, each component of f is a power series in $x_1, \ldots, x_n$, the components of x.*
4. *Suppose $y \in \overline{U}$ and P_y is a matrix such that $|P_y| = 1$ and*

$$P_y[f_x(y)]P_y^{-1} = J,$$

   where J is the real canonical form of $f_x(y)$. Then there exists $M > 0$ such that

$$\operatorname*{lub}_{y \in U} |P_y^{-1}| < M.$$

5. *For each $y \in \overline{U}$, the eigenvalues of the matrix*

$$f_x(y)$$

   satisfy the following hypotheses:

   (*i*) *There exists $r > 0$ such that, for all $y \in U$, the matrix*

$$f_x(y)$$

   has $(n-1)$ eigenvalues that have real parts that are less than or equal to $-r$. If any of the eigenvalues of a matrix $f_x(y)$ has a nonsimple elementary divisor, then we require that $r > M$.

   (*ii*) *The imaginary parts of the eigenvalues of $f_x(y)$ for all $y \in \overline{U}$ have absolute value b, where*

$$0 \leq b < \frac{r - \zeta}{M}.$$

and ζ is a positive number if all the eigenvalues have simple elementary divisors. If any of the eigenvalues have nonsimple elementary divisors, then the imaginary parts of the eigenvalues have absolute value less than or equal to

$$\frac{r - M - \eta}{M} = \frac{r}{M} - 1 - \frac{\eta}{M},$$

where η is a positive number. [Remember that according to (i) if there exists an eigenvalue with a nonsimple elementary divisor, then $r > M$ so that if η is sufficiently small,

$$\frac{r}{M} - 1 - \frac{\eta}{M} > 0.$$

We note that computable criteria for condition 5 are given by Marden (1966), especially on pp. 197 and 203.]

6. There exists $\beta > 0$ such that if $\tilde{x} \in \overline{U}$ and the nth eigenvalue of $f_x(\tilde{x})$ has real part greater than $-r$ and $u(\tilde{x})$ is an eigenvalue of unit length associated with the nth eigenvalue of $f_x(\tilde{x})$, and if $f_{(1)}(\tilde{x})$ is the coefficient of $u(\tilde{x})$ in the expansion of $f(\tilde{x})$ in terms of the n (generalized) eigenvectors of $f_x(\tilde{x})$, then

$$\left| f_{(1)}(\tilde{x}) \right| \geq \beta.$$

7. Let $-q_1, \ldots, -q_k$ denote the real parts of the $(n-1)$ eigenvalues that have real part less than or equal to $-r$. Then for $j = 2, \ldots, k$,

$$|q_1 - q_j| < \zeta.$$

Conclusion: If $x(t)$ is a solution of (5.16) and there exists t_0 such that $x(t_0) \in U$, then $x(t)$ is phase asymptotically stable. If $\overline{U}$ is connected there is exactly one phase asymptotically stable periodic solution $u(t)$ such that

$$O[u(t)] \subset \overline{U}$$

and if $x(t)$ is a solution such that $x(t_0) \in \overline{U}$ for some t_0, then

$$\Omega[x(t)] = O[u(t)].$$

Proof. See Cronin (1980a).

We have already seen that physiologically significant solutions are bounded and we have indicated how boundedness of solutions is established by showing that there is an appropriate bounded set into which all the physiologically meaningful solutions enter and remain. In some cases, it may be possible then to apply Theorem 8 or some theorem like it. This direction is still largely unexplored territory, but there is an extreme case that should be pointed out explicitly.

Suppose that we consider a model of a physiological situation

$$\frac{dx}{dt} = f(x),$$

and suppose there exists a set $\overline{U}$, the closure of a bounded open set, such that all the physiologically meaningful solutions of the model enter $\overline{U}$ and remain thereafter in $\overline{U}$. The next step would be to search for phase asymptotically stable solutions in $\overline{U}$. Since any such phase asymptotically stable solution would be bounded, then Sell's theorem would be immediately applicable to it.

Now suppose that by some means we were able to show that there are no phase asymptotically stable solutions in the set $\overline{U}$. Then it follows that no solution of the model can be used to predict the behavior of the physiological system. That is, if the physiological system is "moving along" a solution and a disturbance "kicks" the system off that solution and onto another, then there is no reason for the system to return to the first solution because the first solution has no stability properties. As pointed out earlier, in mathematical models of physiological systems, we must assume that small disturbances, not taken into account in the model, are constantly impinging on the physiological system. Thus, since the system is kicked from one solution to another in random fashion, the solution of the model cannot be used to predict the behavior of the physiological system. The only prediction that can be made is that the physiological system stays in or near the set $\overline{U}$.

In such a case, there are two possible conclusions. One is, very simply, that the model is a poor description of the physiological system and a better model should be sought. There is a second possibility, which is very serious because it brings into question what can be expected of a mathematical model. Suppose, in the hypothetical case we just described, there is good reason to believe that the model is a fairly accurate description. Then the information that the model gives us is that the physiological system behaves in a random or unpredictable manner. Now it turns out that such behavior or activity is actually observed in some systems. For such cases, a mathematical model that is a system of ordinary differential equations cannot be expected to give much information [For a detailed discussion of a problem of this kind, see Cronin (1977).]

The possibility is related to the subject of "chaotic" systems that has been studied extensively in the last few years and in which many questions remain open.

5.3   Periodic solutions
5.3.1   Autonomous systems

We have already seen in Sell's theorem that the study of asymptotically stable solutions leads to the existence of periodic solutions. The problem of determining whether there exist periodic solutions is an important one for the study of physiological models because, as has been shown in Chapters 3 and 4, we are often concerned with oscillatory phenomena in physiological systems and these are often well described mathematically by periodic solutions. The most conventional approach to the study of periodic solutions is to establish the existence of the periodic solution and then determine whether it is stable. Our next step is to list criteria for the existence of periodic solutions. We start with consideration of the two-dimensional case for which there is an old and very well-known result, the Poincaré–Bendixson theorem.

5.3.1.1   *Poincaré–Bendixson theorem*

Poincaré–Bendixson theorem. *Given the autonomous system*

$$\frac{dx}{dt} = P(x, y),$$

$$\frac{dy}{dt} = Q(x, y),$$

(5.17)

where P, Q are continuous and satisfy a local Lipschitz condition at each point of an open set in R^2, suppose that the solution $S = (x(t), y(t))$ of (5.17) is defined for all $t \geq t_0$, where t_0 is a fixed

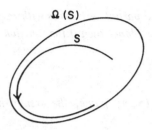

Figure 5.1.

value, and is bounded. Suppose also that $\Omega(S)$ contains no equilibrium points of (5.17). Then one of the following two alternatives holds.

1. Solution S is a periodic solution [in which case $\Omega(S)$ is the orbit of S].
2. $\Omega(S)$ is the orbit of a periodic solution and solution S approaches $\Omega(S)$ spirally from the outside of spirally from the inside (see Fig. 5.1).

The Poincaré–Bendixson theorem suffers from the same usual drawback of pure existence theorems. That is, the existence of a periodic solution is established, but the periodic solution may not be unique, and the theorem gives no hint about how to compute the periodic solution. The Poincaré–Bendixson theorem has another serious limitation. It is well known from consideration of higher-dimensional examples that the theorem is not true if the dimension $n > 2$. Since most of the models we have considered have dimension at least 4, we certainly need a higher-dimensional result.

5.3.1.2 Bendixson criterion

A simple and often convenient test for the nonexistence of periodic solutions in the two-dimensional case is given by the following condition.

Bendixson criterion. *Given the system*

$$\frac{dx}{dt} = P(x, y),$$

$$\frac{dy}{dt} = Q(x, y),$$

where P and Q have continuous first partial derivatives with respect to x and y at each point of the (x, y) plane, then if the function

$$\frac{\partial P}{\partial x} + \frac{\partial Q}{\partial y}$$

is nonzero at each point of the (x, y) plane, the system has no nontrivial periodic solutions.

Proof. See Cronin (1980b) or any standard textbook on differential equations.

5.3.1.3 *Hopf bifurcation theorem*

Once we consider dimension $n > 2$, there are very few results about the existence of periodic solutions of autonomous systems except Sell's theorem. There is, however, one well-known result, the Hopf bifurcation theorem, that should be described in detail.

The Hopf theorem is concerned with an autonomous system with a parameter. Very roughly speaking, it states that periodic solutions may bifurcate (or appear) from an equilibrium point as the parameter is varied through a critical value. The precise statement of the theorem is as follows.

Hopf bifurcation theorem. *Consider the n-dimensional equation*

$$\frac{dx}{dt} = A(\varepsilon)x + f(x, \varepsilon) \tag{5.18}$$

where ε is a parameter, the function f has continuous first derivatives in x and ε, and f is such that

$$\lim_{|x| \to 0} \frac{|f(x, \varepsilon)|}{|x|} = 0$$

uniformly in ε for $|\varepsilon|$ sufficiently small; $A(\varepsilon)$ is a differentiable matrix function of ε. Let $x(t, x^0, \varepsilon)$ denote the solution of (5.18) such that $x(0, x^0, \varepsilon) = x^0$. Suppose that matrix $A(\varepsilon)$ has the eigenvalues $\alpha(\varepsilon) + i\beta(\varepsilon)$ and $\alpha(\varepsilon) - i\beta(\varepsilon)$, where $\alpha(\varepsilon)$ is a real differentiable function such that $\alpha(0) = 0$, $\alpha'(0) \neq 0$, and $\beta(0) \neq 0$. Suppose also that $i\beta(0)$ is an eigenvalue of multiplicity 1 of the matrix

$A(0)$ and suppose that $A(0)$ has no eigenvalue of the form $i[n\beta(0)]$, where $n = 0, \pm 2, \pm 3, \ldots$. Then there is an interval $I = (-r, r)$, where $r > 0$, and there exist real-valued differentiable functions $\varepsilon(s), h(s), c_3(s), \ldots, c_n(s)$, all with domain I, such that

$$\varepsilon(0) = h(0) = c_3(0) = \cdots = c_n(0) = 0$$

and such that, if

$$c(s) = [0, s, c_3(s), \ldots, c_n(s)],$$

then the solution

$$x[t, c(s), \varepsilon(s)]$$

of (5.18) has period

$$\frac{2\pi}{\beta(0)} [1 + h(s)].$$

The function $\varepsilon(s)$ is nonpositive or nonnegative, so the periodic solution is obtained only for $\varepsilon > 0$ or $\varepsilon < 0$. Thus the periodic solution "bifurcates" from the equilibrium point 0 as ε passes through the value 0.

It is clearly important to determine if the periodic solution thus obtained is stable, and there are explicit, that is, computable, criteria for stability. A good discussion of these is given by Poore (1976). The Hopf theorem is directly applicable only to a limited class of equations. Probably its primary importance lies in showing that a certain kind of qualitative behavior (the appearance of periodic solutions) can occur and this kind of qualitative behavior is analogous to or serves as a model of experimentally observed behavior of certain biological and hydrodynamical systems.

Essentially we have just two results about the existence of periodic solutions if $n > 2$: Sell's theorem and the Hopf bifurcation theorem. The Hopf theorem is a local result: It says that if a parameter is changed slightly, then under certain circumstances a small periodic solution appears near an equilibrium point. Sell's theorem is an in-the-large result: It says that if there is a phase asymptotically stable solution, that solution approaches an equilibrium point or a periodic solution.

5.3.2 Periodic solutions of equations with a periodic forcing term
5.3.2.1 *The need for such periodic solutions*

The theorems about periodic solutions of autonomous systems that we have described are likely to be useful in the study of our physiological models because most of these models are autonomous systems. However, there are important modifications of these models that are systems of differential equations that have periodic forcing terms. To study these modifications we need theorems about the existence of periodic solutions of nonautonomous systems, and this is the topic to which we turn now.

Before embarking on the mathematics, we describe briefly a few of these modifications that are nonautonomous systems. We have already described (in Chapter 4) the example of a periodic current stimulus applied to a squid axon. This was studied numerically by Berkinblit et al. (1970), but a rigorous qualitative study requires the use of theorems about periodic solutions of nonlinear nonautonomous systems. Much more important examples are nonautonomous modifications of the model of the Purkinje fiber. If the action of a Purkinje fiber in its natural setting is to be studied, then the influence of the periodic pacemaker pulse, that is, the pulse originating in the pacemaker region of the sino-atrial node, must be taken into account. Mathematically, this means adding a periodic forcing term to the model of the Purkinje fiber, for example, the Noble equations or the McAllister–Noble–Tsien equations.

5.3.2.2 *A general theorem about existence of periodic solutions*

We consider an n-dimensional nonautonomous system of the form

$$\frac{dx}{dt} = f(t, x)$$

or

$$\frac{dx_1}{dt} = f_1(t, x_1, \ldots, x_n)$$
$$\vdots \tag{5.19}$$
$$\frac{dx_n}{dt} = f_n(t, x_1, \ldots, x_n),$$

where $\partial f_i / \partial t$, $\partial f_i / \partial x_j$ $(i, j = 1, \ldots, n)$ exist and are continuous in

R^{n+1} and $f(t, x)$, regarded as a function of t, has period T, where T is a positive constant; that is, for all x and all t,

$$f(t, x) = f(t + T, x).$$

The question is: Does equation (5.19) have a solution of period T? In order to state a theorem that answers this question, we let $x(t, c)$ denote the solution of (5.19) such that $x(0, c) = c$.

Theorem 9. *Suppose there is a bounded open set $U \subset R^n$ such that $\bar{U}$ is homeomorphic to a closed ball in R^n and such that $\bar{U}$ has the following property: if $c \in \bar{U}$, then $x(T, c) \in \bar{U}$. (This property can be described in words as: a solution that passes through a point in $\bar{U}$ is also in $\bar{U}$ after the time interval T has elapsed.)*
Conclusion: There is a point $c_0 \in \bar{U}$ such that $x(t, c_0)$ has period T.

This theorem is proved by applying the Brouwer fixed point theorem to the mapping M of $\bar{U}$ into itself, defined as follows: If $c \in \bar{U}$,

$$M: c \to x(T, c).$$

The hypothesis of the theorem guarantees that mapping M takes U into itself and, hence, by the Brouwer fixed point theorem, mapping M has a fixed point c_0, that is,

$$x(T, c_0) = c_0 = x(0, c_0).$$

It is easy to show that this condition implies that the solution $x(t, c_0)$ has period T.

One drawback of this theorem is that the Brouwer fixed point theorem is used and consequently the proof of the theorem yields no information whatever about how to compute the solution unless we refer to a constructive proof of the Brouwer theorem. Second, since we do not known how to compute solution $x(t, c_0)$, there is little hope of investigating its stability. Finally, the theorem just says that at least one periodic solution exists. We have no estimate of an upper bound for the number of periodic solutions. There might, indeed, exist an infinite set of periodic solutions. In order to circumvent this difficulty, we impose additional hypotheses on $f_j(t, x_1, \ldots, x_n)$. Let us suppose that each $f_j(t, x_1, \ldots, x_n)$ has the form

$$P_j(t, x_1, \ldots, x_n),$$

where $P_j(t, x_1,\ldots, x_n)$ is a polynomial of degree q_j in $x_1,\ldots, x_n$ and the coefficients in P_j are continuous functions of t that have period T except for the coefficients of terms of degree q_j. These last coefficients are assumed to be constants. Thus, $P_j(x_1,\ldots, x_n)$ may be written as

$$P_j(t, x_1,\ldots, x_n) = R_j(x_1,\ldots, x_n) + Q_j(t, x_1,\ldots, x_n),$$

where R_j is a polynomial homogeneous of degree q_j in $x_1,\ldots, x_n$, $Q_j(t, x_1,\ldots, x_n)$ is a polynomial of degree $q_j - 1$ in $x_1,\ldots, x_n$, and the coefficients are continuous functions of t with period T. Finally we assume that the homogeneous polynomials $R_1(x_1,\ldots, x_n),\ldots, R_n(x_1,\ldots, x_n)$ have no common zeros in R^n except the origin. [A necessary and sufficient condition that this hold is that the resultant of $R_1,\ldots, R_n$ is nonzero; see van der Waerden (1940) or Macauley (1916).] Under these hypotheses, we have the following result.

Theorem 10. *The number m of periodic solutions of* (5.19) *is either infinite or, if m is finite,*

$$1 \le m \le \prod_{j=1}^{n} q_j.$$

If $n = 2$, then m is finite and $1 \le m \le q_1 q_2$. For arbitrary n, if we add an arbitrarily small constant term of f, m becomes finite and $1 \le m \le \prod_{j=1}^{n} q_j$.

Proof. See Cronin (1980b, Appendix).

5.3.2.3 Branching of periodic solutions

Just as the Hopf bifurcation theorem shows that a certain kind of qualitative behavior (the appearance of periodic solutions) can occur in autonomous equations, certain theorems about branching of periodic solutions show that analogous qualitative behavior occurs for nonautonomous equations. We consider the classical problem of searching for periodic solutions of an n-dimensional system of the form

$$\frac{dx}{dt} = f(t, x, \varepsilon), \tag{5.20}$$

where ε is a real parameter such that $|\varepsilon|$ is small and f has components $f_j(t, x_1, \ldots, x_n, \varepsilon)$, $j = 1, \ldots, n$, where each f_j has continuous partial derivatives of the third order with respect to $t, x_1, \ldots, x_n, \varepsilon$. Further, we assume that for each fixed value of ε, $f_j(t, x_1, \ldots, x_n, \varepsilon)$ has period $T(\varepsilon)$ where $T(\varepsilon)$ is a differentiable function of ε. We assume that for $\varepsilon = 0$, (5.20) has a solution $x(t)$ of period $T(0)$ and we study the following problem.

Problem 1. If $|\varepsilon|$ is sufficiently small, does (5.20) have a solution $x(t, \varepsilon)$ of period $T(\varepsilon)$ such that for each real t,

$$\lim_{\varepsilon \to 0} x(t, \varepsilon) = x(t)?$$

By using a strategic change of variables and applying Floquet theory [see Cronin (1980b)], this problem can be reduced to the study of the equation

$$\frac{dx}{dt} = Ax + \varepsilon F(t, x, \varepsilon) + G(t), \qquad (5.21)$$

where A is a constant real matrix, the functions F and G have continuous first derivatives in all variables, and F and G have period T in t, where T denotes the number $T(0)$. Problem 1 can be rephrased as:

Problem 2. If $|\varepsilon|$ is sufficiently small, does equation (5.21) have solutions of period T?

To answer this question, we have first a classical theorem due to Poincaré.

Theorem 11. *If the equation*

$$\frac{dx}{dt} = Ax$$

has no nontrivial solutions of period $2n\pi/T$ *(or, equivalently, if matrix* A *has no eigenvalues of the form* $2n\pi i/T$, *where* $n = 0, \pm 1, \pm 2, \ldots$*), then there exist numbers* $\eta_1 > 0, \eta_2 > 0$ *such that for each* ε *with* $|\varepsilon| < \eta_1$, *there is a unique vector* $c = c(\varepsilon)$ *such that*

$$|c(\varepsilon) - c_0| < \eta_2,$$

where

$$c_0 = -(e^{TA} - I)^{-1} e^{TA} \int_0^T e^{-sA} G(s) \, ds$$

and

$$x(t, \varepsilon, c(\varepsilon))$$

is a solution of (5.21) *that has period T.*

Now suppose that matrix A does have one or more eigenvalues of the form $2n\pi i/T$. This is sometimes called the resonance case. In certain applications, the resonance case is by far the more important case, but its analysis requires a more detailed study of the influence of the nonlinear terms. For an account of analytical methods of study, see Hale (1963). For a discussion of the use of topological methods, see Cronin (1964, 1980b). These mathematical results can be used to describe such oscillatory phenomena as entrainment of frequency and subharmonic oscillations. Topological methods can be used to obtain lower and upper bounds on the number of periodic solutions and information (albeit limited) about the stability of these solutions.

5.4 Singularly perturbed equations
5.4.1 Introduction

It has long been known that problems in electrical circuit theory often give rise to so-called singularly perturbed systems of differential equations [see, e.g., Mishchenko and Rozov (1980, Chap. I)]. We consider a singularly perturbed system of the form

$$\varepsilon \frac{dx_1}{dt} = f_1(x_1, \ldots, x_n)$$

$$\vdots$$

$$\varepsilon \frac{dx_k}{dt} = f_k(x_1, \ldots, x_n),$$

$$\frac{dx_{k+1}}{dt} = f_{k+1}(x_1, \ldots, x_n)$$

$$\vdots$$

$$\frac{dx_n}{dt} = f_n(x_1, \ldots, x_n).$$

That is, some (but not all) of the first derivatives are multiplied by a small parameter ε. The study, both qualitative and quantitative, of solutions of such singularly perturbed systems is by no means complete, but we shall summarize some of the important results of these studies because there is considerable evidence to indicate that many of the models of electrically active cells are singularly perturbed systems of ordinary differential equations or can be approximated by such systems. We will review this evidence at the beginning of Chapter 6.

5.4.2 Some examples
5.4.2.1 *Two-dimensional examples*
We begin by looking at a fairly simple two-dimensional system

$$\frac{dx}{dt} = \frac{1}{\varepsilon}(y - x^3),$$
$$\frac{dy}{dt} = g(x, y),$$

(5.22)

where $g(x, y)$ has continuous first partial derivatives with respect to x and y at each point of R^2. To get an idea of how the solutions of this system behave when ε is very small and positive, we notice that if $y = x^3$, then $dx/dt = 0$; but at any point of R^2 that is not on the curve $y = x^3$, $|dx/dt|$ is very large if ε is sufficiently small. That means that except for points on the curve $y = x^3$ or points very close to that curve, the vector $(dx/dt, dy/dt)$ is almost horizontal, pointing left or right depending on the sign of dx/dt. Thus as a rough sketch of the vector field we obtain Fig. 5.2.

In order to investigate what happens very close to the curve $y = x^3$, it is necessary to specify $g(x, y)$ and to study its values on the curve $y = x^3$. For the present, it is sufficient to observe that if a

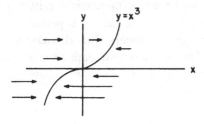

Figure 5.2.

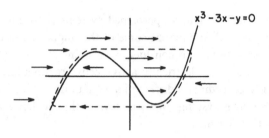

Figure 5.3.

solution of (5.22) is on or very near the curve $y = x^3$, then it will stay close to the curve $y = x^3$. The curve $y = x^3$ is sometimes called the slow manifold because it is only on or very near this curve that the motion along the solution curves is not very fast in a nearly horizontal direction. Without looking more closely at this configuration, let us consider a slightly more complicated singularly perturbed system:

$$\frac{dx}{dt} = -\frac{1}{\varepsilon}(x^3 - 3x - y),$$

$$\frac{dy}{dt} = g(x, y),$$

(5.23)

where $g(x, y)$ is as in (5.22). A rough sketch of the vector field is shown in Fig. 5.3.

Now further, let us suppose that in (5.22), $g(x, y)$ is positive on the half-plane $y < 0$ and that $g(x, y)$ is negative on the half-plane $y > 0$. This, together with Fig. 5.2, suggests that all the solutions of (5.22) approach the equilibrium point $(0, 0)$. Next suppose that in (5.23), $g(x, y)$ is positive on the half-plane $x < 0$ and negative on the half-plane $x > 0$. What happens to the solutions of (5.23) is not clear. However, it seems possible that there might exist a solution close to the dashed curve in Fig. 5.3, that is, a periodic solution consisting of two "fast" path segments that alternate with two "slow" path segments. The fast and slow path segments are determined by "solving" the degenerate system, that is, system (5.23), with $\varepsilon = 0$:

$$y = x^3 - 3x,$$

$$\frac{dy}{dt} = g(x, y).$$

(Notice that we multiply the first equation of (5.23) by ε before setting $\varepsilon = 0$.) The "fast segments" are $\{ y = 2, \ -1 \le x \le \sqrt{3} \}$ and $\{ y = -2, \ -\sqrt{3} \le x \le 1 \}$. The slow segments are then $\{(x, y)/-\sqrt{3} \le x \le -1, \ y = x^3 - 3x\}$ and $\{(x, y)/1 \le x \le \sqrt{3}, \ y = x^3 - 3x\}$. Now in fact, this sketchily stated conjecture is true, that is, it can be proved in very general circumstances that if we find a "discontinuous solution" of the degenerate system (i.e., find a connected sequence of fast and slow segments), then for ε small and positive, the singularly perturbed system has a solution that is very close to the discontinuous solution of the degenerate system. This solution need not, of course, be periodic. We have represented that particular case in Figure 5.3 because it is a familiar and important case. Equally important for our purposes is the following kind of example. Consider

$$\frac{dx}{dt} = -\frac{1}{\varepsilon}(x^3 - 3x + y),$$

$$\frac{dy}{dt} = g(x, y),$$

where $g(x, y) = 0$ is the straight line indicated in Fig. 5.4 and the signs of $g(x, y)$ are as indicated in Fig. 5.4. It is easy to see that a rough sketch of the vector field is as indicated in Fig. 5.4. The point E is an equilibrium point and a possible discontinuous solution starts at the point A and is sketched with a dashed curve. Note that it approaches the equilibrium point. Also note that a

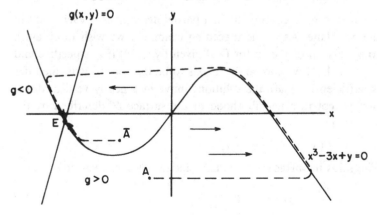

Figure 5.4.

solution of the degenerate system that starts at the point $\bar{A}$ and is also sketched with a dashed curve moves much more directly to the point E and that the points A and $\bar{A}$ may be chosen close together even though the solutions that pass through them behave very differently.

5.4.2.2 A three-dimensional example

We consider one more example, a three-dimensional singularly perturbed system. We give this example partly because it is enlightening to illustrate the ideas that have been advanced with another and significantly different example and partly because the three-dimensional example is an illustration of the elementary catastrophe that is most widely used in applications of catastrophe theory. See Zeeman (1972, 1973, 1977). (It should be noted that Zeeman has proposed in his papers a three-dimensional model of the voltage-clamp behavior of the squid axon. Because Zeeman's model has serious deficiencies [see Cronin (1981)], we have not included a description of it in Chapter 3.)

We consider the singularly perturbed system

$$\frac{dx}{dt} = -\frac{1}{\varepsilon}\left(x^3 + xy + z\right),$$

$$\frac{dy}{dt} = f(y, z), \tag{5.24}$$

$$\frac{dz}{dt} = g(y, z),$$

where f, g have continuous first partial derivatives at each point in the yz plane. As in the preceding examples, we wish to obtain a rough sketch of the vector field given by (5.24) if ε is positive and very small. If we choose the x axis to be the vertical axis, then if ε is sufficiently small, the solutions move in a nearly vertical direction except in a neighborhood of the surface $\mathscr{S}$ described by the equation

$$F(x, y, z) = x^3 + xy + z = 0.$$

The folds in surface $\mathscr{S}$ are characterized by the equations

$$\frac{\partial F}{\partial x} = 3x^2 + y = 0 \tag{5.25}$$

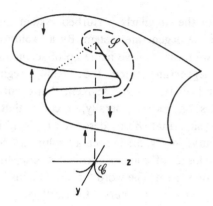

Figure 5.5.

and

$$F(x, y, z) = x^3 + xy + z = 0, \qquad (5.26)$$

and the projections of the fold curves onto the (y, z) plane, which are derived by eliminating x between (5.25) and (5.26), are described by

$$4y^3 + 27z^2 = 0. \qquad (5.27)$$

Equation (5.27) describes a cusp $\mathscr{C}$ and it is easy, by elementary arguments, to show that the surface $\mathscr{S}$ and the cusp $\mathscr{C}$ appear as sketched in Fig. 5.5. (Figure 5.5 is a representation of the elementary catastrophe that is most widely used in applications of catastrophe theory.) Let us assume that functions f and g in (5.24) are such that the dashed curve is a solution of the degenerate system. The same kind of arguments used for system (5.23) suggest that system (5.24) has a solution near the discontinuous solution sketched by a dashed curve.

5.4.3 Some theory of singularly perturbed systems
5.4.3.1 *Theorems of Mishchenko and Rozov*
The remainder of our discussion of singularly perturbed systems consists in describing rigorously and precisely the mathematics that is suggested by the preceding examples. In this, we will follow Mishchenko and Rozov (1980), Levinson (1951), and Sibuya (1960). We will state several general theorems concerning the

existence of solutions of the singularly perturbed system that are close to "solutions" of the degenerate system. By a "solution" of the degenerate system we mean a curve made up of segments that are solutions of the degenerate system (i.e., "slow" segments) alternating with the kind of straight line segment that occurred in the preceding examples or, more generally, a curve that is a solution of the "fast" system that will be described in detail in the following. The closed curve in Fig. 5.3 is such a "solution." Such a curve (not necessarily closed, of course) is called a *discontinuous solution* of the degenerate system. The word "discontinuous" refers to the behavior of the tangent of the curve. Generally the tangent changes discontinuously at certain transition points from one curve segment to another.

Now we summarize results of Mishchenko and Rozov (1980). Our first step is to give a formal definition in the *n*-dimensional case of a discontinuous solution of the degenerate solution. For this purpose we need some preliminary definitions. We consider the system

$$\varepsilon \frac{dx^i}{dt} = f^i(x^1, \ldots, x^k, y^1, \ldots, y^l),$$

$$\frac{dy^j}{dt} = g^j(x^1, \ldots, x^k, y^1, \ldots, y^l), \tag{5.28}$$

where $i = 1, \ldots, k$, $j = 1, \ldots, l$, $k + l = n$, and ε is a small positive parameter. The notation

$$x = (x^1, \ldots, x^k),$$

$$f = (f^1, \ldots, f^k),$$

$$y = (y^1, \ldots, y^l),$$

$$g = (g^1, \ldots, g^l),$$

will be used; the *k*-dimensional Euclidean space of points $(x^1, \ldots, x^k)$ is denoted by X^k, the *l*-dimensional Euclidean space of points $(y^1, \ldots, y^l)$ is denoted by Y^l, and the *n*-dimensional Euclidean space of points $(x^1, \ldots, x^k, y^1, \ldots, y^l)$, which is the direct sum of X^k and Y^l, is denoted by R^n. R^n is the domain of the vector functions f and g, and f and g are assumed to have continuous first derivatives in all variables at each point of R^n (so that the standard existence and uniqueness results hold) and also all other derivatives required in the discussion that follows.

The *degenerate system* corresponding to (5.28) is the system obtained from (5.28) by setting $\varepsilon = 0$, that is, the system

$$f^i(x^1,\ldots,x^k,y^1,\ldots,y^l) = 0, \qquad i = 1,\ldots,k,$$

$$\frac{dy^j}{dt} = g^j(x^1,\ldots,x^k,y^1,\ldots,y^l), \qquad j = 1,\ldots,l. \tag{5.29}$$

The *fast-motion equation system* corresponding to (5.28) is the k-dimensional system

$$\varepsilon \frac{dx^i}{dt} = f^i(x^1,\ldots,x^k,y^1,\ldots,y^l), \qquad i = 1,\ldots,k, \tag{5.30}$$

where the numbers $y^1,\ldots,y^l$ are regarded as parameters.

Assumption 1. If $y^1,\ldots,y^l$ are arbitrary fixed values, then each solution $(x^1(t),\ldots,x^k(t))$ of (5.30) is such that if it has an ω-limit set, the ω-limit set is an equilibrium point.

Let Γ denote the l-dimensional surface in R^n defined by the k equations

$$f^i(x^1,\ldots,x^k,y^1,\ldots,y^l) = 0, \qquad i = 1,\ldots,k. \tag{5.31}$$

[In referring to Γ as an l-dimensional surface, we are assuming that the (5.31) are such that Γ is an l manifold.]

Next let $\mathscr{A}(x^1,\ldots,x^k,y^1,\ldots,y^l)$ be the $k \times k$ matrix defined by

$$\mathscr{A}(x^1,\ldots,x^k,y^1,\ldots,y^l) = \left[\frac{\partial f^q}{\partial x^r}(x^1,\ldots,x^k,y^1,\ldots,y^l)\right],$$

$$q, r = 1,\ldots,k.$$

The matrix $\mathscr{A}(x^1,\ldots,x^k,y^1,\ldots,y^l)$ is defined at all points of R^n. Let

$$\Gamma_- = \big\{(x^1,\ldots,x^k,y^1,\ldots,y^l) \in \Gamma / \text{all eigenvalues of}$$

$$\mathscr{A}(x^1,\ldots,x^k,y^1,\ldots,y^l) \text{ have negative real parts}\big\}$$

The set Γ_- is called the *stable region* of Γ.

Assumption 2. No point of Γ_- is an equilibrium point of (5.28). That is, if $(\bar{x}^1,\ldots,\bar{x}^k,\bar{y}^1,\ldots,\bar{y}^l) \in \Gamma_-$, then

$$\left(f^1(x^1,\ldots,\bar{x}^k,\bar{y}^1,\ldots,\bar{y}^l),\ldots,f^k(\bar{x}^1,\ldots,\bar{x}^k,\bar{y}^1,\ldots,\bar{y}^l), \right.$$
$$\left. g^1(\bar{x}^1,\ldots,\bar{x}^k,\bar{y}^1,\ldots,\bar{y}^l),\ldots,g^l(\bar{x}^1,\ldots,\bar{x}^k,\bar{y}^1,\ldots,\bar{y}^l) \right)$$
$$\neq (0,\ldots,0,0,\ldots,0).$$

Let

$$\Gamma_0 = \left\{ (x^1,\ldots,x^k,y^1,\ldots,y^l) \right.$$
$$\left. \in \Gamma / \det \mathscr{A}(x^1,\ldots,x^k,y^1,\ldots,y^l) = 0 \right\}.$$

The points in Γ_0 are called *nonregular points*. In general, the set Γ_0 is an $(l-1)$-dimensional subset of Γ and

$$\Gamma - \Gamma_0 = C \cup D,$$

where

$$C = \left\{ (\tilde{x}^1,\ldots,\tilde{x}^k,\tilde{y}^1,\ldots,\tilde{y}^l) \right.$$
$$\left. \in \Gamma / \det \mathscr{A}(\tilde{x}^1,\ldots,\tilde{x}^k,\tilde{y}^1,\ldots,\tilde{y}^l) > 0 \right\},$$
$$D = \left\{ (\underline{x}^1,\ldots,\underline{x}^k,\underline{y}^1,\ldots,\underline{y}^l) \right.$$
$$\left. \in \Gamma / \det \mathscr{A}(\underline{x}^1,\ldots,\underline{x}^k,\underline{y}^1,\ldots,\underline{y}^l) < 0 \right\}.$$

From the definition of Γ_0, it follows that if $(x_0^1,\ldots,x_0^k,y_0^1,\ldots,y_0^l) \in \Gamma_0$, then the matrix

$$\mathscr{A}(x_0^1,\ldots,x_0^k,y_0^1,\ldots,y_0^l)$$

has at least one eigenvalue that is zero.

Definition 1. A nonregular point $S(x_0^1,\ldots,x_0^k,y_0^1,\ldots,y_0^l)$ is a *junction point* if and only if the following conditions are satisfied.

1. S is not an equilibrium point of (5.28). Since $S \in \Gamma_0 \subset \Gamma$, then $f(S) = 0$. Hence, this condition means that $g(S) \neq 0$.
2. Zero is an eigenvalue of algebraic multiplicity 1 of $\mathscr{A}(S)$. All the other eigenvalues of $\mathscr{A}(S)$ have negative real parts.
3. The fast-motion equation system with $(y^1,\ldots,y^l) = (y_0^1,\ldots,y_0^l)$, that is, the system

$$\varepsilon \frac{dx^i}{dt} = f^i(x^1,\ldots,x^k,y_0^1,\ldots,y_0^l), \qquad (5.32)$$

has the point $(x_0^1, \ldots, x_0^k)$ as an equilibrium point [since $f(S) = 0$]. We require that $(x_0^1, \ldots, x_0^k)$ be an isolated equilibrium point and that $(x_0^1, \ldots, x_0^k)$ be the α-limit set of exactly one orbit of (5.32). That is, there is exactly one orbit of (5.32), the underlying point set of a solution $x(t)$ of (5.32), such that $\lim_{t \to -\infty} x(t) = (x_0^1, \ldots, x_0^k)$.

4. Let $X_{y_0^1, \ldots, y_0^l}^k$ denote the k-dimensional hyperplane

$$\left\{ (x^1, \ldots, x^k, y^1, \ldots, y^l) \mid y^1 = y_0^1, \ldots, y^l = y_0^l \right\}.$$

Then, if $X_{y_0^1, \ldots, y_0^l}^k$ is translated by any sufficiently small vector, it does not contain an equilibrium point of the fast-motion equation system near S.

Definition 2. Let $(\tilde{x}^1, \ldots, \tilde{x}^k, \tilde{y}^1, \ldots, \tilde{y}^l) \in R^n - \Gamma$ and consider the orbit of

$$\varepsilon \frac{dx^i}{dt} = f^i(x^1, \ldots, x^k, \tilde{y}^1, \ldots, \tilde{y}^k), \tag{5.33}$$

which passes through the point $(\tilde{x}^1, \ldots, \tilde{x}^l)$. If the ω-limit set of this orbit is a stable equilibrium point $(\tilde{\tilde{x}}^1, \ldots, \tilde{\tilde{x}}^k)$ and if

$$\left(\tilde{\tilde{x}}^1, \ldots, \tilde{\tilde{x}}^k, \tilde{y}^1, \ldots, \tilde{y}^l \right) \in \Gamma_-,$$

then the point $(\tilde{\tilde{x}}^1, \ldots, \tilde{\tilde{x}}^k, \tilde{y}^1, \ldots, \tilde{y}^l)$ is called the *drop point corresponding to the point* $(\tilde{x}^1, \ldots, \tilde{x}^k, \tilde{y}^1, \ldots, \tilde{y}^l)$.

Definition 3. Let $S(x_0^1, \ldots, x_0^k, y_0^1, \ldots, y_0^l)$ be a junction point that satisfies the following condition: The orbit of

$$\varepsilon \frac{dx^i}{dt} = f^i(x^1, \ldots, x^k, y_0^1, \ldots, y_0^l),$$

which has $(x_0^1, \ldots, x_0^k)$ as its α-limit set (condition 3 in Definition 1), has the property that its ω-limit set is a stable equilibrium point $P(x_*^1, \ldots, x_*^k, y_0^1, \ldots, y_0^l)$ and

$$P\left(x_*^1, \ldots, x_*^k, y_0^1, \ldots, y_0^l \right) \in \Gamma_-.$$

Then P is the *drop point following the junction point S*.

With Assumptions 1 and 2 and Definitions 1–3, we are ready to define a discontinuous solution of the degenerate system (5.29). In describing the discontinuous solution, we will use the term "phase point" and speak of the "phase point traversing" a curve. The term "phase point" simply refers to a position on a curve and how the phase point "traverses" the curve is given by the analytic description of the curve. A discontinuous solution of (5.29) with initial point $Q \in R^n$ is the continuous curve $\mathcal{T} \subset R^n$ obtained by applying successively the following steps.

Step 1. If $Q(\tilde{x}^1,\ldots,\tilde{x}^k,\tilde{y}^1,\ldots,\tilde{y}^l) \notin \Gamma$, then the system

$$\varepsilon \frac{dx^i}{dt} = f^i(x^1,\ldots,x^k,\tilde{y}^1,\ldots,\tilde{y}^l), \qquad i=1,\ldots,k,$$

has a unique solution

$$(x^1(t),\ldots,x^k(t))$$

such that

$$x^i(t_0) = \tilde{x}^i, \qquad i=1,\ldots,k.$$

We assume that the ω-limit set of the solution

$$(x^1(t),\ldots,x^k(t))$$

is a stable equilibrium point $(\tilde{\tilde{x}}^1,\ldots,\tilde{\tilde{x}}^k)$ and that

$$(\tilde{\tilde{x}}^1,\ldots,\tilde{\tilde{x}}^k,\tilde{y}^1,\ldots,\tilde{y}^l) \in \Gamma_-.$$

That is, we assume that $(\tilde{\tilde{x}}^1,\ldots,\tilde{\tilde{x}}^k,\tilde{y}^1,\ldots,\tilde{y}^l)$ is a drop point corresponding to Q. The first section of the continuous curve $\mathcal{T}$ is the curve

$$\{(x^1(t),\ldots,x^k(t),\tilde{y}^1,\ldots,\tilde{y}^l) | t \geq t_0\}.$$

The phase point traverses this section of $\mathcal{T}$ instantaneously and this is called a fast-motion part of $\mathcal{T}$.

Step 2. If $Q(\tilde{\tilde{x}}^1,\ldots,\tilde{\tilde{x}}^k,\tilde{y}^1,\ldots,\tilde{y}^l) \in \Gamma_-$, then from the definition of Γ_- and by the implicit function theorem, the system of equations

$$f^i(x^1,\ldots,x^k,y^1,\ldots,y^l) = 0, \qquad i=1,\ldots,k,$$

can be solved uniquely for $x^1,\ldots,x^k$ in terms of $y^1,\ldots,y^l$ in a neighborhood of the point $Q(\tilde{\tilde{x}}^1,\ldots,\tilde{\tilde{x}}^k,\tilde{y}^1,\ldots,\tilde{y}^l)$. That is, there exist differentiable functions

$$x^i(y^1,\ldots,y^l), \qquad i=1,\ldots,k,$$

such that

$$x^i(\tilde{y}^1,\ldots,\tilde{y}^l)=\tilde{\tilde{x}}^i$$

and such that

$$f^i\big[x^1(y^1,\ldots,y^l),\ldots,x^k(y^1,\ldots,y^l),y^1,\ldots,y^l\big]=0.$$
$$(5.34)$$

Substituting

$$x^i=x^i(y^1,\ldots,y^l), \qquad i=1,\ldots,k,$$

into (5.29), we obtain (5.34) and

$$\frac{dy^j}{dt}=g^j\big[x^1(y^1,\ldots,y^l),\ldots,x^k(y^1,\ldots,y^l),y^1,\ldots,y^l\big],$$
$$j=1,\ldots,l. \quad (5.35)$$

The existence theorem can be applied to (5.35) and we obtain the conclusion that there exists a solution

$$(y^1(t),\ldots,y^l(t))$$

of (5.35) such that

$$y^j(t_0)=\tilde{y}^j, \qquad j=1,\ldots,l,$$

and such that

$$(x^1[y^1(t),\ldots,y^l(t)],\ldots,x^k[y^1(t),\ldots,y^l(t)],$$
$$y^1(t),\ldots,y^l(t))\in\Gamma.$$

This last statement follows from (5.34). Next we assume that there exists $t_1>t_0$ such that the solution

$$(y^1(t),\ldots,y^l(t))$$

is defined for all $t\in[t_0,t_1)$ and that

$$\lim_{t\to t_1}(y^1(t),\ldots,y^l(t))$$

exists and that

$$S=\lim_{t\to t_1}(x^1[y^1(t),\ldots,y^l(t)],\ldots,x^k[y^1(t),\ldots,y^l(t)],$$
$$y^1(t),\ldots,y^l(t))$$

is a junction point in Γ_0. In this case, the resulting section of the curve $\mathcal{T}$ is

$$\{(x^1[y^1(t_1),\ldots,y^l(t_1)],\ldots,x^k[y^1(t_1),\ldots,y^l(t)],$$
$$y^1(t),\ldots,y^l(t))/t\in[t_0,t_1]\}.$$

The time interval during which the phase point moves along this section is $t_1 - t_0$ and this section is called a slow-motion part of $\mathcal{T}$.

Step 3. If $Q(x_0^1,\ldots,x_0^k,y_0^1,\ldots,y_0^l)$ is a junction point, we assume that there exists a drop point P following the junction point Q. Then the corresponding section of the curve $\mathcal{T}$ is defined as follows. Let the orbit of

$$\varepsilon\frac{dx^i}{dt}=f^i(x^1,\ldots,x^k,y_0^1,\ldots,y_0^l),\qquad i=1,\ldots,k,$$

which has $(x_0^1,\ldots,x_0^k)$ as its α-limit point, be described by the solution

$$(x^1(t),\ldots,x^k(t))$$

The corresponding sections of $\mathcal{T}$ is

$$\{(x^1(t),\ldots,x^k(t),y_0^1,\ldots,y_0^l)|t\text{ real}\}\cup\{P\},$$

where

$$P=\lim_{t\to\infty}(x^1(t),\ldots,x^k(t),y_0^1,\ldots,y_0^l).$$

The phase point traverses this section of $\mathcal{T}$ instantaneously and such a section is called a fast-motion part of $\mathcal{T}$.

We illustrate these definitions with an example (shown in Fig. 5.6)

$$\frac{dx}{dt}=\frac{-1}{\varepsilon}(x^3-3x-y),$$
$$\frac{dy}{dt}=-\left(x-\frac{1}{2}\right).$$

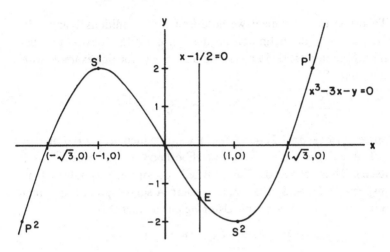

Figure 5.6.

The degenerate system is

$$y = x^3 - 3x,$$

$$\frac{dy}{dt} = -\left(x - \frac{1}{2}\right).$$

The fast-motion equation system is

$$\varepsilon \frac{dx}{dt} = -\left(x^3 - 3x - y\right).$$

For fixed y, this is a single first-order differential equation and hence Assumption 1 is automatically satisfied. The set Γ is the curve

$$y = x^3 - 3x.$$

The matrix $\mathscr{A}$ is the 1×1 matrix (i.e., just a scalar)

$$\mathscr{A} = -3x^2 + 3.$$

The set Γ_- is

$$\Gamma_- = \left\{(x, y) \mid y = x^3 - 3x \text{ and } x^2 > 1\right\}.$$

Since E is the only equilibrium point, Assumption 2 is certainly satisfied. The set Γ_0 consists of the two points $S^1(-1, 2)$ and $S^2(1, -2)$. Also each of the points S^1 and S^2 is a junction point.

To prove this statement we note first that conditions 1 and 2 in Definition 1 (the definition of junction point) are obviously satisfied. Equation (5.32) in condition 3 becomes, for this example with the point S^1,

$$\varepsilon \frac{dx}{dt} = -(x^3 - 3x - 2) = 2 + 3x - x^3.$$

Since $-(x^3 - 3x - 2) = -(x - 2)(x + 1)^2$, then $dx/dt = 0$ at $x = -1$ and $dx/dt > 0$ for x in a neighborhood of -1 (except at -1 itself). Thus, condition 3 is satisfied. Condition 4 is satisfied by inspection. Thus, S^1 is a junction point. A similar proof shows that S^2 is also a junction point. Using again the fact that

$$-(x^3 - 3x - 2) = -(x - 2)(x + 1)^2,$$

we see easily that $P^1(2,2)$ is a drop point following S^1 and P^2 $(-2, -2)$ is a drop point following S^2.

Now we are ready to describe some discontinuous solutions. Suppose first that the initial point Q is such that $Q \in \Gamma_-$; in particular, take Q to be, say, the point $(\sqrt{3}, 0)$. By Step 2 in the definition of discontinuous solution, the first section of the discontinuous solution is

$$\{(x, y) | 1 \le x \le \sqrt{3}, \, y = x^3 - 3x\}.$$

Since S^2 is a junction point, then by Step 3 in the definition of discontinuous solution, the second section of the discontinuous solution is

$$\{(x, y) | -2 \le x \le 1, \, y = -2\}.$$

The sections that follow are

$$\{(x, y) | -2 \le x \le -1, \, y = x^3 - 3x\},$$
$$\{(x, y) | -1 \le x \le 2, \, y = 2\},$$

and

$$\{(x, y) | \sqrt{3} \le x \le 2, \, y = x^3 - 3x\}.$$

The discontinuous solution is a simple closed curve.

We obtain a second discontinuous solution by considering an initial point Q such that $Q \notin \Gamma$. In particular, take Q to be, say, the point $(1, 0)$. Then by using the steps in the definition of discontinuous solutions, we see easily that the discontinuous solution that has as its initial point $Q = (1, 0)$ consists of the sections

$$\{(x, y)|1 \le x \le \sqrt{3}, \, y = 0\}$$

and then the same sections as for the preceding discontinuous solution.

For a three-dimensional example, see Mishchenko and Rozov (1980, p. 19).

The first mathematical question that we consider is the following. Suppose that $\mathscr{T}_0$ is a discontinuous solution with initial point $Q_0 = (x_0^1, \ldots, x_0^k, y_0^1, \ldots, y_0^l)$. Let $\mathscr{T}_\varepsilon$ be the orbit of the solution of (5.28) that has Q_0 as its initial value at $t = t_0$. That is, $\mathscr{T}_\varepsilon$ is the orbit of the solution $(x^1(t), \ldots, x^k(t), y^1(t), \ldots, y^l(t))$ of (5.28) such that $(x^1(t_0), \ldots, x^k(t_0), y^1(t_0), \ldots, y^l(t_0)) = Q_0(x_0^1, \ldots, x_0^k, y_0^1, \ldots, y_0^l)$. How far apart are $\mathscr{T}_0$ and $\mathscr{T}_\varepsilon$, especially as $\varepsilon \to 0$? The first answer to this question is given by Mishchenko and Rozov (1980, p. 174, Theorem 1).

Theorem 12 (Mishchenko and Rozov). *Let $\bar{\mathscr{T}}_0$ be a finite segment defined on the interval $[t_0, t_1]$ of the discontinuous solution $\mathscr{T}_0$. That is, $\bar{\mathscr{T}}_0$ consists of the slow-motion parts of $\mathscr{T}_0$ whose domains are contained in $[t_0, t_1]$ and the fast-motion parts that occur between these slow-motion sections. Then if ε is sufficiently small, the solution*

$$\left(x^1(t), \ldots, x^k(t), y^1(t), \ldots, y^l(t)\right),$$

of which $\mathscr{T}_\varepsilon$ is the orbit, is defined for all $t \in [t_0, t_1]$. If

$$\bar{\mathscr{T}}_\varepsilon = \left\{\left(x^1(t), \ldots, x^k(t), y^1(t), \ldots, y^l(t)\right)|t \in [t_0, t_1]\right\}$$

and if

$$\rho\left(\bar{\mathscr{T}}_0, \bar{\mathscr{T}}_\varepsilon\right) = \operatorname*{lub}_{p \in \bar{\mathscr{T}}_\varepsilon} d\left[p, \bar{\mathscr{T}}_0\right],$$

where

$$d\left[p, \bar{\mathscr{T}}_0\right] = \operatorname*{glb}_{q \in \bar{\mathscr{T}}_0} |p - q|,$$

then

$$\lim_{\varepsilon \to 0} \rho\left(\bar{\mathscr{T}}_0, \bar{\mathscr{T}}_\varepsilon\right) = 0.$$

Roughly speaking, this theorem says that if ε is small enough, then $\bar{\mathscr{T}}_0$ is a good approximation to the orbit $\bar{\mathscr{T}}_\varepsilon$. This is an extremely powerful result because, as we shall see in later applications, finding a discontinuous solution is, in some cases, much easier than directly studying the solutions of the singularly perturbed system.

The second question that we want to consider is the following. Suppose that the discontinuous solution $\mathscr{T}_0$ is a closed curve. Then if ε is small, is $\mathscr{T}_\varepsilon$ a closed curve? In other words, if ε is small, does the singularly perturbed system have a periodic solution whose orbit is near $\mathscr{T}_0$? This question is answered by Mishchenko and Rozov (1980, p. 204, Theorem 1).

We impose the same hypotheses as in the preceding theorem except that we assume also that $\mathscr{T}_0$ is a simple closed curve and that $\mathscr{T}_0$ is isolated and stable. That $\mathscr{T}_0$ is isolated and stable means the following. If S is a junction point of $\mathscr{T}_0$ and if $N(S)$ is any sufficiently small neighborhood in Γ_0 of S, let $p \in N(S) - S$. We consider the discontinuous solution with orbit $\tilde{\mathscr{T}}_0$ which contains p. (It is, of course, necessary to prove that such a discontinuous solution exists.) Assume that if we follow $\tilde{\mathscr{T}}_0$ after p, $\tilde{\mathscr{T}}_0$ will intersect $N(S)$ in a point $\tilde{p}$. Define the mapping

$$\phi: p \to \tilde{p},$$

that is,

$$\tilde{p} = \phi(p).$$

That $\mathscr{T}_0$ is isolated and stable means that we assume that ϕ has a unique fixed point S and that the linear part of ϕ is a mapping from $N(S)$ into $N(S)$ and that this mapping also has a unique fixed point, the point S.

We need just one more assumption, which is a nondegeneracy condition. Without loss of generality, we may assume that $\mathscr{T}_0$ consists of four parts: two slow-motion parts and two fast-motion parts. Let S_1 and S_2 be the two junction points of $\mathscr{T}_0$. Let S_1^* be the $(l-1)$-dimensional plane tangent to Γ_0 at S_1, and let S_2^* be the $(l-1)$-dimensional plane tangent to Γ_0 at S_2. We make the following assumption. [Nondegeneracy assumption; see Mishchenko and Rozov (1980, p. 202).]

The vector $g(S_1)$ is transversal to S_1^* (i.e., if v_1 is a nonzero vector in S_1^*, then $(|g(S_1) \cdot v_1|)/(|g(S_1)| |v_1|) < 1$. Also the vector $g(S_1)$ is transversal to S_2^*.

Now we have:

Theorem 13 [Mishchenko and Rozov (1980, p. 204)]. *There exists* $\varepsilon_0 > 0$ *such that if* $0 < \varepsilon < \varepsilon_0$, *then* (5.28) *has a closed trajectory* $\mathcal{T}_\varepsilon$, *which describes a periodic solution, such that*

$$\lim_{\varepsilon \to 0} \rho(\mathcal{T}_0, \mathcal{T}_\varepsilon) = 0.$$

[*Whether this closed trajectory* $\mathcal{T}_\varepsilon$ *is unique is not known; see Mishchenko and Rozov* (1980, *p.* 203).]

For the special case $n = 2$, Theorem 13 has a simpler stronger form, which we will use in the analysis of the FitzHugh–Nagumo equations.

Theorem 14 [Mishchenko and Rozov (1980, pp. 141–142)]. *Given the singularly perturbed system*

$$\varepsilon \frac{dx}{dt} = f(x, y),$$

$$\frac{dy}{dt} = g(x, y),$$

(5.36)

where x, y *are scalars, we assume that the following conditions are satisfied*:

1. f, g have continuous second derivatives of all kinds at each point in the xy plane.
2. If Γ denotes the curve defined by the equation

   $$f(x, y) = 0$$

   then each point of Γ is an ordinary point, that is, at each point of Γ,

   $$[f_x(x, y)]^2 + [f_y(x, y)]^2 > 0.$$

3. The nonregular points of Γ, that is, the points at which

   $$f_x(x, y) = 0,$$

   are isolated and each of them is nondegenerate, that is, at each of them,

   $$f_{xx}(x, y) \neq 0.$$

4. Let the regular part of Γ between nonregular points p and $\bar{p}$ be described by

$$x = x_0(y), \qquad y_p \leq y \leq y_{\bar{p}}.$$

If for all $y \in [y_p, y_{\bar{p}}]$, we have

$$f_x(x_0(y), y) < 0,$$

then this regular part of Γ is called a *stable part*. We assume that $g(x, y) \neq 0$ at all points of the stable part and at all nonregular points of Γ. [In other words, we assume that there is no equilibrium point on a stable part and that no nonregular point is an equilibrium point of (5.36).]
5. No two nonregular points of Γ have the same ordinates.
6. Suppose the degenerate system corresponding to (5.36), that is, the system

$$0 = f(x, y),$$

$$\frac{dy}{dt} = g(x, y),$$

has a closed discontinuous solution $\mathcal{T}_0$.

Conclusion: *If ε is sufficiently small, the system* (5.36) *has a unique stable limit cycle $\mathcal{T}_\varepsilon$ such that*

$$\lim_{\varepsilon \to 0} \rho(\mathcal{T}_0, \mathcal{T}_\varepsilon) = 0.$$

5.4.3.2 *Theorems of Levinson*

We will see in Chapter 6 that the preceding theorems from Mishchenko and Rozov are useful in the analysis of our models of electrically active cells. However, there is one important class of solutions that cannot be obtained by using the Mishchenko–Rozov theory. If we consider the FitzHugh–Nagumo equations or the Hodgkin–Huxley equations and regard them as singularly perturbed systems, then a solution that describes an action potential is not a periodic solution but is a solution that approaches a stable equilibrium point. This stable equilibrium point describes the resting or stable quiescent condition of the unstimulated axon. Now if such a solution is close to a discontinuous solution, then this must be a discontinuous solution that approaches a stable equilibrium point, that is, the kind of discontinuous solution indicated in Fig. 5.4. But the equilibrium point E in Fig. 5.4 is an asymptotically stable equilibrium point. Thus, Assumption 2, which holds in Theorems 12 and 13, is violated. Also condition 4 in the statement of Theorem 14 is violated. Thus, Theorems 12, 13, and 14 are not applicable. However, certain parts of the theory due to Levinson (1951) are applicable. Although we shall use only a two-dimensional version of Levinson's results, we summarize some of the n-dimensional theory so as to indicate the relationship and overlap of the Levinson theory and the Mishchenko–Rosov theory. It should be noted also that Levinson's approach to singularly perturbed equations is somewhat more general than that of Mishchenko and Rozov because Levinson includes nonautonomous as well as autonomous equations, whereas Mishchenko and Rozov consider only autonomous equations. We consider the system

$$\frac{dx_i}{dt} = \left[f_i(x_1,\ldots,,x_n, u, t, \varepsilon) \right] \frac{du}{dt} + \phi_i(x_1,\ldots, x_n, u, t, \varepsilon),$$

$$i = 1,\ldots, n,$$

$$\varepsilon \frac{d^2 u}{dt^2} + \left[g(x_1,\ldots, x_n, u, t, \varepsilon) \right] \frac{du}{dt}$$

$$+ h(x_1,\ldots, x_n, u, t, \varepsilon) = 0, \tag{5.37}$$

where $x_1,\ldots, x_n, u, t, \varepsilon$ are real and $f_1,\ldots, f_n, \phi_1,\ldots, \phi_n, g, h$ are real-valued continuous functions for $x_1,\ldots, x_n, u, t$ real and $\varepsilon \geq 0$. Most frequently we will write (5.37) in vector form as

$$\frac{dx}{dt} = f \frac{du}{dt} + \phi,$$

$$\varepsilon \frac{d^2 u}{dt^2} + g \frac{du}{dt} + h = 0. \tag{5.38}$$

The *degenerate system* of (5.38) is

$$\frac{dy}{dt} = f \frac{dv}{dt} + \phi,$$

$$g \frac{dv}{dt} + h = 0. \tag{5.39}$$

The degenerate system is obtained from (5.38) by setting ε equal to 0 and using the notation (y, v) in place of (x, u). As will be seen later, the change of notation is a clarifying convenience.

System (5.38) includes as a special case the system

$$
\frac{dx}{dt} = H(x, w, t, \varepsilon),
$$

$$
\varepsilon \frac{dw}{dt} = G(x, w, t, \varepsilon),
$$

(5.40)

where x, H are n vectors and G, w are scalars. [System (5.40) is the kind of system that occurs in mathematical modelling of electrically active cells.] To show that (5.40) is a special case of (5.38), we argue as follows. First, if the right-hand sides of (5.40) are linear in w, let u be such that

$$
\frac{du}{dt} = w.
$$

Then (5.40) can be written as

$$
\frac{dx}{dt} = f(x, t, \varepsilon) \frac{du}{dt} + \phi(x, t, \varepsilon),
$$

$$
\varepsilon \frac{d^2 u}{dt^2} + g(x, t, \varepsilon) \frac{du}{dt} + h(x, t, \varepsilon) = 0.
$$

(5.41)

System (5.41) is a special case of (5.38). Second, in the general case, that is, the case in which at least one of the right-hand sides of (5.40) is not linear in w, by differentiating the last equation in (5.40) with respect to t, we obtain

$$
\frac{dx}{dt} = H(x, w, t, \varepsilon),
$$

$$
\varepsilon \frac{d^2 w}{dt^2} = \frac{\partial G}{\partial w} \frac{dw}{dt} + \frac{\partial G}{\partial x} \frac{dx}{dt} + \frac{\partial G}{\partial t}.
$$

(5.40′)

This last system is a special case of (5.38) with $f \equiv 0$.

Our next step is to define a discontinuous solution of (5.39). This step is parallel to the similar step taken in the Mishchenko–Rozov theory. However, in Levinson's work, the concept of discontinuous solution differs significantly from the concept of discontinuous solution used by Mishchenko and Rozov. At the outset there is a large difference because system (5.39) is, in general, nonautonomous. Hence, the discontinuous solution will be a curve in (y, v, t) space, that is, $(n + 2)$ space. [Since Mishchenko and Rozov deal only with an autonomous system, their discontinuous solutions are curves in (y, v) space, i.e., $(n + 1)$ space.]

In order to define a discontinuous solution, we shall be obliged to write a fairly lengthy account that will include making a number of assumptions. The clearest way to present these assumptions is to state them as they are required rather than to try to list all of them in advance. We shall indicate the introduction of an assumption simply by using the word "assume" in italics.

First we *assume* that there exists a point $A = (\bar{y}, \bar{v}, \alpha)$ in (y, v, t) space such that

$$g(\bar{y}, \bar{v}, \alpha) > 0.$$

Then in a neighborhood of $(\bar{y}, \bar{v}, \alpha)$, system (5.39) becomes a conventional $(n + 1)$-dimensional system of ordinary differential equations

$$\frac{dy}{dt} = f\frac{dv}{dt} + \phi,$$
$$\frac{dv}{dt} = -\frac{h}{g}$$

or

$$\frac{dy}{dt} = f\left[-\frac{h}{g}\right] + \phi,$$
$$\frac{dv}{dt} = -\frac{h}{g}. \tag{5.42}$$

Consequently, it is reasonable to expect that we can solve (5.39) in a neighborhood of $(\bar{y}, \bar{v}, \alpha)$ or, more precisely, that there exists a solution $(y(t), v(t))$ of (5.39) that satisfies the initial value

$$y(\alpha) = \bar{y},$$
$$v(\alpha) = \bar{v}.$$

Indeed, if f, g, h, ϕ satisfy slightly stronger conditions than just continuity, then the existence theorem stated earlier in this chapter guarantees the existence of such a solution. We shall *assume* that there exists a number τ_1 with the following properties:

1. $\tau_1 > \alpha.$
2. There exists a solution $(y(t), v(t))$ of (5.42) such that

$$(y(\alpha), v(\alpha)) = (\bar{y}, \bar{v})$$

   and the domain of $(y(t), v(t))$ contains $[\alpha, \tau_1)$. We will denote this solution by S_0. Also $\lim_{t \uparrow \tau_1} y(t) = y(\tau_1 - 0)$ and $\lim_{t \uparrow \tau_1} v(t) = v(\tau_1 - 0)$ exist and are finite. We will denote $y(\tau_1 - 0)$ and $v(\tau_1 - 0)$ by y_B, v_B.
3. If $t \in [\alpha, \tau_1)$,

$$g[y(t), v(t), t, 0] > 0$$

   and

$$\lim_{t \uparrow \tau_1} g[y(t), v(t), t, 0] = 0.$$

Let B_1 denote the point (y_B, v_B, τ_1) in (y, v, t)-space and let AB_1 denote the union of the orbit of solution S_0 for $\alpha \le t < \tau_1$ and the point B_1. That is,

$$AB_1 = \{(y(t), v(t), t) | t \in [\alpha, \tau_1)\} \cup \{B_1\}.$$

The first arc of the discontinuous solution is defined to be AB_1.

The second arc of the discontinuous solution is a curve in the hyperplane $t = \tau_1$ in (y, v, t) space. In order to describe this second arc, we first make another assumption. Let

$$I(y, v, t) = \sum_{i=1}^{n} \frac{\partial g}{\partial y_i} f_i + \frac{\partial g}{\partial v}$$

and denote $I(y_B, v_B, \tau_1)$ by I_B. We *assume* that $I_B \ne 0$. We also *assume* that $h(y_B, v_B, \tau_1, 0) \ne 0$. The second arc of the discontinuous solution is the orbit of the solution of

$$\frac{dy}{dv} = f(y, v, \tau_1, 0) \tag{5.43}$$

with initial value

$$y = y_B = \lim_{t \uparrow \tau_1} y(t),$$

$$v = v_B = \lim_{t \uparrow \tau_1} v(t).$$

We arrive at this description of the second arc by the following considerations. Since $h(y_B, v_B, \tau_1, 0) \neq 0$, then if $t < \tau_1$ and if $\tau_1 - t$ is sufficiently small, it follows that

$$-\frac{h[y(t), v(t), t, 0]}{g[y(t), v(t), t, 0]} \neq 0$$

and hence $dv/dt \neq 0$. Thus, dt/dv is defined. Multiplying both sides of the equations in (5.39) by dt/dv, we get

$$\frac{dy}{dt}\frac{dt}{dv} = f\frac{dv}{dt}\frac{dt}{dv} + \phi\frac{dt}{dv},$$

$$g\frac{dv}{dt}\frac{dt}{dv} + h\frac{dt}{dv} = 0$$

or

$$\frac{dy}{dv} = f + \phi\frac{dt}{dv},$$

$$g + h\frac{dt}{dv} = 0. \tag{5.44}$$

If t is held constant so that $dt/dv = 0$, then (5.44) becomes

$$\frac{dy}{dv} = f[y(t), v(t), t, 0],$$

$$g[y(t), v(t), t, 0] = 0. \tag{5.45}$$

Since

$$\lim_{t \uparrow \tau_1} g[y(t), v(t), t, 0] = 0,$$

then as $t \uparrow \tau_1$ system (5.45) becomes

$$\frac{dy}{dv} = f[y(\tau_1 - 0), v(\tau_1 - 0), \tau_1, 0].$$

It is these considerations that suggest that the second arc of the discontinuous solution is the orbit of the solution of

$$\frac{dy}{dv} = f(y, v, \tau_1, 0) \tag{5.43}$$

with initial value

$$y = y_B,$$
$$v = v_B.$$

In (5.43), variable v is an independent variable. We will consider v to be increasing above v_B or decreasing below v_B according as $h[y_B, v_B, \tau_1, 0]$ is negative or positive. (Remember that we have assumed that $h[y_B, v_B, \tau_1, 0] \neq 0$.) The reason for this consideration is that if $t < \tau_1$,

$$\frac{dv}{dt} = -\frac{h}{g}.$$

Hence, if $\tau_1 - t$ is sufficiently small, then since

$$g[y(t), v(t), t, 0] > 0,$$

it follows that dv/dt has the sign opposite to the sign of $h[y_B, v_B, \tau_1, 0]$. For definiteness, we will assume that $h[y_B, v_B, \tau_1, 0]$ is negative and hence that v is increasing.

Let $y(v)$ be the solution of (5.43) such that

$$y(v_B) = y_B.$$

By assumption

$$g[y_B, v_B, \tau_1, 0] = 0. \tag{5.46}$$

But if $v \neq v_B$ and $|v - v_B|$ is sufficiently small, then

$$\frac{d}{dv} g[y(v), v, \tau_1, 0] = \sum_{i=1}^{n} \frac{\partial g}{\partial y_i} \frac{dy_i}{dv} + \frac{\partial g}{\partial v}.$$

Thus

$$\lim_{v \to v_B} \frac{d}{dv} g[y(v), v, \tau_1, 0] = \sum_{i=1}^{n} \frac{\partial g}{\partial y_i} f_i(y_B, v_B, \tau_1, 0)$$

$$+ \frac{\partial g}{\partial v}(y_B, v_B, \tau_1, 0).$$

But this last expression is I_B, which has been assumed to be nonzero. This fact along with (5.46) implies that if $|v - v_B|$ is nonzero but sufficiently small, it follows that

$$g[y(v), v, \tau_1, 0] \neq 0.$$

Hence, if $[v - v_B]$ is positive but sufficiently small,

$$\int_{v_B}^{v} g[y(v), v, \tau_1, 0] \, dv \neq 0.$$

Now we *assume* that there is a smallest number v_c such that $v_c > v_B$, the domain of solution $y(v)$ of (5.43) contains the interval $[v_B, v_c]$, and

$$\int_{v_B}^{v_c} g[y(v), v, \tau_1, 0] \, dv = 0.$$

Let C_1 denote the point $(y(v_c), v_c, \tau_1)$. Denote $g[y(v_c), v_c, \tau_1, 0]$ by g_c and *assume* that

$$g_c > 0.$$

(This implies that I_B, which was earlier assumed to be nonzero, is actually negative.) Since $g_c \neq 0$, then in a neighborhood of the point C_1, the degenerate system (5.39) is a conventional $(n + 1)$-dimensional system

$$\frac{dy}{dt} = f\left[-\frac{h}{g}\right] + \phi,$$

$$\frac{dv}{dt} = -\frac{h}{g}.$$

Consequently it is reasonable to *assume* that there exists a solution $(y(t), v(t))$ such that

$$(y(\tau_1) \, v(\tau_1)) = (y(v_c), v_c).$$

We *assume* that there exists $\tau_2 > \tau_1$ such that:

1. The domain of $(y(t), v(t))$ contains the set $[\tau_1, \tau_2)$.
2. For $t \in [\tau_1, \tau_2)$,

$$g[y(t), v(t), t, 0] > 0.$$

3. $\lim_{t \uparrow \tau_2} y(t) = y(\tau_2 - 0)$ and $\lim_{t \uparrow \tau_2} v(t) = v(\tau_2 - 0)$ exist and are finite,
4. $\lim_{t \uparrow \tau_2} g[y(t), v(t), t, 0] = 0$.

Let B_2 denote the point $(y(\tau_2 - 0), v(\tau_2 - 0), \tau_2)$ in (y, v, t) space. Next we *assume* that the same kind of hypotheses hold for the point B_2 that held for B_1. Hence, we can then proceed in the same way at B_2 as we did at B_1 with v increasing or decreasing according as $h[y(\tau_2 - 0), v(\tau_2 - 0), \tau_2, 0]$ is negative or positive.

Continuing in this manner, we obtain a curve

$$AB_1C_1B_2 \cdots B_NC_NA'$$

(where A' is a point at which $g \neq 0$).

Definition. The curve

$$AB_1C_1B_2 \cdots B_NC_NA',$$

which will be denoted by $\mathscr{S}$, is a *discontinuous solution* of (5.39).

Now we are ready to state Levinson's theorem. We make the following assumptions:

A1. There exists a discontinuous solution $\mathscr{S}$ of (5.39) defined for $\alpha \le t \le \beta$.

A2. There exists an open set R contained in (y, v, t) space such that:

(i) $\mathscr{S} \subset R$.

(ii) $f(y, v, t, \varepsilon), \phi, g, h$ and their first partial derivatives with respect to $y_1, \ldots, y_n, v, t$ are uniformly continuous and bounded as functions of y, v, t, ε for $(y, v, t) \in R$ and $0 \le \varepsilon \le r$, where r is a small positive number.

(iii) g, h have continuous second partial derivatives on $R \times [0, r]$. [Levinson (1951, p. 75) points out that this assumption can be avoided.]

In the statement of the theorem, we use the vector norm that was introduced earlier in this chapter. That is, if x is the n vector $(x_1, \ldots, x_n)$, then the norm of x is

$$|x| = \sum_{i=1}^{n} |x_i|.$$

Theorem 15 (Levinson's theorem). *Suppose that* (5.39) *satisfies the preceding assumptions A1 and A2. If $\varepsilon, \delta_1, \delta_2$ are sufficiently small and positive, there exists a solution $(x(t), u(t))$ of* (5.38) *with a domain that contains $[\alpha, \beta]$ for any set of initial values $(x(\alpha), u(\alpha))$ that satisfy the inequalities*

$$|x(\alpha) - y(\alpha)| + |u(\alpha) - v(\alpha)| \le \delta_1,$$

$$\left| \frac{du}{dt}(\alpha) - \frac{dv}{dt}(\alpha) \right| \le \frac{\delta_2}{\varepsilon}.$$

As $\varepsilon, \delta_1, \delta_2 \to 0$, *the curve in* (x, u, t) *space described by* $(x(t), u(t), t)$ *for* $\alpha \leq t \leq \beta$ *approaches* $\mathscr{S}$. *In particular, for any fixed* $\delta > 0$, *the function*

$$|x(t) - y(t)| + |u(t) - v(t)|$$

converges uniformly to zero on the set

$$E = [\alpha + \delta, \tau_1 - \delta] \cup [\tau_1 + \delta, \tau_2 - \delta] \cdots [\tau_N + \delta, \beta]$$

as $\varepsilon, \delta_1, \delta_2 \to 0$. *Also the functions*

$$\frac{du}{dt} - \frac{dv}{dt} \quad \text{and} \quad \frac{d^2u}{dt^2} - \frac{d^2v}{dt^2}$$

converge uniformly to zero on the set E *as* $\varepsilon, \delta_1, \delta_2 \to 0$.

Theorem 15 was introduced by Levinson to treat a system of the form (5.37). Actually, our models are all of the special form (5.40). Consequently when Theorem 15 is applied, the initial values are specified values only of $w(\alpha)$ and $x(\alpha)$. The $dw/dt(\alpha)$ is then specified by the second equation in system (5.40). [If we specify a value of $dw/dt(\alpha)$, then, integrating the second equation of (5.40'), we would obtain

$$\varepsilon \frac{dw}{dt} = G(x, w, t, \varepsilon) + C,$$

where C is a constant of integration whose value would be determined by the specified value of $dw/dt(\alpha)$. Thus, unless $C = 0$, we would not have the system (5.40) that we are trying to study.] Thus δ_2, and the inequality

$$\left| \frac{dw}{dt}(\alpha) - \frac{dv}{dt}(\alpha) \right| \leq \frac{\delta_2}{\varepsilon}$$

can be disregarded; that is, the version of Theorem 15 that we will use is:

Theorem 15R (revised). *Suppose that the degenerate system of* (5.40) *satisfies assumptions A1 and A2. Then if* ε *and* δ_1 *are sufficiently small, and positive, there is a solution* $(w(t), x(t))$ *of* (5.40) *with the solution defined for* $\alpha \leq t \leq \beta$ *for any set of initial values* $w(\alpha), x(\alpha)$ *such that*

$$|w(\alpha) - v(\alpha)| + |x(\alpha) - y(\alpha)| \leq \delta_1.$$

[The discontinuous solution given by assumption A1 is described by $(y(t), v(t))$.] Also as ε, δ_1 tend to zero, the curve representing the solution $(w(t), x(t))$ in (w, x, t) space tends to $\mathscr{S}$. In particular, for any fixed $\delta > 0$,

$$|w(t) - v(t)| + |x(t) - y(t)|$$

tends uniformly to zero over the intervals

$$\alpha \le t \le \tau_1 - \delta,$$

$$\tau_1 + \delta \le t \le \tau_2 - \delta,$$

$$\tau_N + \delta \le t \le \beta,$$

as $\varepsilon, \delta_1 \to 0$.

If Theorem 15 is to be applied to system (5.40), the first step is to differentiate the equation

$$\varepsilon \frac{dw}{dt} = G(x, w, t, \varepsilon)$$

with respect to t. As a consequence of this, it turns out that the discontinuous solution with initial point A, where $\partial G / \partial v \ne 0$, stays (for some time interval) in the manifold in (y, v, t) space described by

$$G(y, v, t, 0) = G_A,$$

where G_A denotes the value of G at the point A. The reason for this is that we have

$$\varepsilon \frac{dw}{dt} = G(x, w, t, \varepsilon)$$

and, after differentiation,

$$\varepsilon \frac{d^2 w}{dt^2} = \frac{d}{dt} [G(x, w, t, \varepsilon)].$$

Setting $\varepsilon = 0$, we get

$$\frac{d}{dt} [G(x, w, t, 0)] = 0.$$

The first step in obtaining the discontinuous solution is to solve the system

$$\frac{dy}{dt} = H(y, w, t, 0),$$

$$\frac{d}{dt}[G(y, w, t, 0)] = 0 \tag{5.47}$$

or

$$\frac{dy}{dt} = H(y, v, t, 0),$$

$$\left(\frac{\partial G}{\partial v}\right)\left(\frac{dv}{dt}\right) + \left(\frac{\partial G}{\partial y}\right)\frac{dy}{dt} + \frac{\partial G}{\partial t} = 0$$

or

$$\frac{dy}{dt} = H(y, v, t, 0),$$

$$\frac{dv}{dt} = -\left(\frac{\partial G}{\partial v}\right)^{-1}\left[\left(\frac{\partial G}{\partial y}\right)\frac{dy}{dt} + \frac{\partial G}{\partial t}\right]. \tag{5.48}$$

We seek a solution $(y(t), v(t))$ such that

$$y(\alpha) = y_A,$$

$$v(\alpha) = v_A,$$

and the point $A = (y_A, v_A, \alpha)$ is such that at A, $\partial G/\partial v \neq 0$. Let $G_A = G(y_A, v_A, \alpha, 0)$. Then

$$G[y(t), v(t), t, 0] = G_A$$

because

$$G[y(\alpha), v(\alpha), \alpha, 0] = G[y_A, v_A, \alpha, 0] = G_A$$

and by (5.48) or, equivalently, (5.47),

$$\frac{dG}{dt}[y(t), v(t), t, 0] = 0$$

so that $G[y(t), v(t), t, 0]$ is a constant. Thus the first arc of the discontinuous solution lies in the manifold

$$G(y, v, t, 0) = G_A.$$

But we seek solutions of system (5.40), the degenerate system of which is

$$\frac{dx}{dt} = H(x, w, t, 0),$$

$$0 = G(x, w, t, 0).$$

Thus, we seek discontinuous solutions, the slow arcs of which are contained in the manifold described by

$$G(x, w, t, 0) = 0.$$

Hence, the point A must be contained in this manifold.

However, Theorem 15 can be extended to a discontinuous solution, the first arc of which is fast, by using arguments in Levinson's paper [Levinson (1951, pp. 81–84, Proof of Theorem 1, Part 2)]. We will need this extension in Chapter 6 for the study of the FitzHugh–Nagumo equations.

We illustrate the preceding theory with the example considered earlier:

$$\frac{dy}{dt} = -\left(x - \frac{1}{2}\right),$$

$$\frac{dx}{dt} = -\frac{1}{\varepsilon}(x^3 - 3x - y). \tag{5.49}$$

To cast this system in the form of (5.37), we differentiate the second equation with respect to t and obtain

$$\frac{dy}{dt} = -\left(x - \frac{1}{2}\right),$$

$$\varepsilon\frac{d^2x}{dt^2} = [-3x^2 + 3]\frac{dx}{dt} + \frac{dy}{dt}$$

or

$$\frac{dy}{dt} = -\left(x - \frac{1}{2}\right),$$

$$\varepsilon\frac{d^2x}{dt^2} = [-3x^2 + 3]\frac{dx}{dt} - \left(x - \frac{1}{2}\right). \tag{5.50}$$

This is a special case of (5.37) with $y = x_1$, $x = u$, and

$f = 0$,

$\phi = \left(x - \frac{1}{2} \right)$,

$g = (3x^2 - 3)$,

$h = \left(x - \frac{1}{2} \right)$.

The corresponding degenerate system is

$$\frac{dy}{dt} = -\left(x - \frac{1}{2} \right),$$

$$(3x^2 - 3)\frac{dx}{dt} = -\left(x - \frac{1}{2} \right). \tag{5.51}$$

The manifold M, described by

$$G(x, w, t, 0) = 0,$$

becomes, in this case,

$$x^3 - 3x - y = 0.$$

(This manifold is, of course, the curve shown in Fig. 5.6.) Let A be a point on this manifold such that

$$g = 3x^2 - 3 > 0.$$

(Any point on M such that $|x| > 1$ satisfies the condition.) We seek a solution of (5.51) that has the initial value A at some value t_0 of t. If we apply Levinson's theory to the degenerate system (5.51), then we consider curves in (y, x, t) space, where these curves are described by solutions of (5.51). However, in this case the equations dealt with are autonomous. Hence, we consider the projections of these curves into (y, x) space. Theorem 15R can be used to conclude that the projections of the curves described by the discontinuous solution and the solution of (5.49) itself stay close together.

Referring to Fig. 5.6, let A be a point on the manifold described by

$$x^3 - 3x - y = 0$$

such that

$$g = 3x^2 - 3 > 0.$$

For example, suppose

$$A = (-\sqrt{3}, 0).$$

As shown earlier, the first arc of the discontinuous solution, which has A as its initial point, lies in the manifold

$$x^3 - 3x - y = 0,$$

and the arc is that part of the manifold that joins the point $A = (-\sqrt{3}, 0)$ and $S^1 = (-1, 2)$. The point $(-1, 2)$ is the first point at which

$$g = 3x^2 - 3$$

is zero. Thus S^1 is the point B_1 described in the definition of the discontinuous solution. [More precisely, S^1 is the projection in (y, x) space of the point B_1.] Now for this example the equation (5.43) becomes

$$\frac{dy}{dx} = 0$$

because in this example the function f is identically zero. Hence, the second arc of the discontinuous solution is a segment of the line

$$y = 2.$$

The segment starts at $(-1, 2)$. Since $h < 0$ at $x = 1$, then x is increasing and the integral

$$\int_{v_B}^{v} g[y(v), v, \tau_1, 0] \, dv$$

for this case becomes

$$\int_{-1}^{x} (3x^2 - 3) \, dx = [x^3 - 3x]_{-1}^{x} = x^3 - 3x + 1 - 3$$

$$= x^3 - 3x - 2$$

$$= (x + 1)^2 (x - 2).$$

Thus, the point v_c is 2 and the second arc of the discontinuous solution is the line segment

$$\{(x, y) | -1 \le x \le 2, \ y = 2\}.$$

The third arc of the discontinuous solution is the subset of manifold

$$x^3 - 3x - y = 0,$$

which is the point set

$$\{(x, y)|1 \leq x \leq 2, \ y = x^3 - 3x\}.$$

We are especially interested in the case of relaxation oscillations, that is, the case in which the discontinuous solution is a closed curve and nearby solutions of the singularly perturbed system are periodic. Although the results of Mishchenko and Rozov deal with this problem, we describe Levinson's somewhat different approach to the problem because it is valuable to have as much information as possible about the question. In order to deal with this case, we will need the following further theorem. First, we denote by $\partial/\partial a$ differentiation with respect to one of the $(n + 1)$ initial coordinates $(w(\alpha), x(\alpha))$ or with respect to one of the corresponding initial coordinates $(v(\alpha), y(\alpha))$.

Theorem 16. *With the same hypotheses as in the preceding theorem, we have, if $\varepsilon, \delta_1, \delta_2 \to 0$,*

$$\left| \frac{\partial w}{\partial a}(t) - \frac{\partial v}{\partial a}(t) \right| + \left| \frac{\partial x}{\partial a}(t) - \frac{\partial y}{\partial a}(t) \right| \to 0,$$

$$\left| \frac{\partial}{\partial a} \frac{dw}{dt} - \frac{\partial}{\partial a} \frac{dv}{dt} \right| \to 0.$$

The convergence is uniform over the same set of intervals as in the preceding theorem. Moreover, if b denotes the initial value of du/dt at α, then

$$\left| \frac{\partial x}{\partial b}(t) \right| + \left| \frac{\partial u}{\partial b}(t) \right| \to 0$$

and

$$\left| \frac{\partial}{\partial b} \frac{du}{dt} \right| \to 0$$

uniformly over the same set of intervals as $\varepsilon, \delta_1, \delta_2 \to 0$.

Now suppose that G and H, as functions of t, have period T where T is a positive constant. That is, suppose that for all x, w, t, ε,

$$G(x, w, t, +T, \varepsilon) = G(x, w, t, \varepsilon),$$

$$H(x, w, t, +T, \varepsilon) = H(x, w, t, \varepsilon).$$

Suppose further that (5.39) has a discontinuous solution with period T, that is, the point A' coincides with point A and the interval (α, β) is $(\alpha, \alpha + T)$. By using the immediately preceding theorem and conventional arguments, we obtain:

Theorem 17. *Let* $D(a_1, \ldots, a_{n+1}, b, \varepsilon)$ *denote the determinant*

$$\begin{vmatrix} \dfrac{\partial x_1}{\partial a_1} - 1 & \dfrac{\partial x_1}{\partial a_2} & \cdots & \dfrac{\partial x_1}{\partial a_{n+1}} & \dfrac{\partial x_1}{\partial b} \\[2ex] \dfrac{\partial x_2}{\partial a_1} & \dfrac{\partial x_2}{\partial a_2} - 1 & \cdots & \dfrac{\partial x_2}{\partial a_{n+1}} & \dfrac{\partial x_2}{\partial b} \\[2ex] & & \vdots & & \\[2ex] \dfrac{\partial w}{\partial a_1} & & \cdots & \dfrac{\partial w}{\partial a_{n+1}} - 1 & \dfrac{\partial w}{\partial b} \\[2ex] \dfrac{\partial}{\partial a_1}\dfrac{dw}{dt} & & \cdots & \dfrac{\partial}{\partial b}\dfrac{dw}{dt} - 1 \end{vmatrix}$$

where the entries in the determinant are evaluated at $t = \beta = \alpha + T$ *and* $a_1, \ldots, a_{n+1}, b$ *denote initial values* $x_1(\alpha), \ldots, x_n(\alpha)$, $w(\alpha)$, $dw/dt(\alpha)$. *Suppose that D is continuous and nonzero for ε sufficiently small, and for the terms* $|a_j - y_j(\alpha)|$ $(j = 1, \ldots, n)$, $|a_{n+1} - v(\alpha)|$, *and* $|b - (G_y H + G_t)/G_v|$, *where* $(G_y H + G_t)/G_v$ *is evaluated at the point A, all sufficiently small. Then for sufficiently small positive ε, equation (5.37) has a unique nearby solution of period T.*

If G, H are independent of t, and (5.39) has a discontinuous solution of period T, and if the hypotheses of the preceding theorem are satisfied, then a similar conclusion holds except that the period of the periodic solution of (5.37) is $T(\varepsilon)$, where $T(\varepsilon)$ is a continuous function of ε and $\lim_{\varepsilon \to 0} T(\varepsilon) = T$.

5.4.3.3 *Theorem of Sibuya*

The work of Levinson is restricted to the case in which just one of the equations is singularly perturbed, that is, w is a scalar (or real variable). Some of Levinson's work (not the part on periodic solutions) has been extended to the general case by Sibuya (1960). It is necessary to use Sibuya's result in order to search for nonperiodic solutions of singularly perturbed systems in which there are at least two equations with a factor $1/\varepsilon$. As we shall see later, this is exactly the problem that must be studied when the Hodgkin–Huxley equations are regarded as a singularly perturbed system.

Before stating Sibuya's results, we point out explicitly some conditions in the results. First we note that the vector norm used by Sibuya is equivalent to, but not identical with, the vector norm used by Levinson. Second, certain characteristic roots of coefficient matrices that occur have value zero. These characteristic roots are required to be simple. (This is a kind of nondegeneracy hypothesis.) Finally, the nonzero characteristic roots are all required to be positive.

The system of equations studied by Sibuya is the following:

$$\varepsilon \frac{d^2 U}{dt^2} + \tilde{A}(t, U, x, \varepsilon) \frac{dU}{dt} + \tilde{a}(t, U, x, \varepsilon) = 0,$$

$$\frac{dx}{dt} = \tilde{B}(t, U, x, \varepsilon) \frac{dU}{dt} + \tilde{b}(t, U, x, \varepsilon),$$

(5.52)

where x and U are m- and m^1-dimensional vectors, respectively, t is a real scalar, and ε is a real nonnegative parameter. It is assumed that there exists an open set R in (t, U, x) space such that the components of the matrices $\tilde{A}$, $\tilde{B}$ and the vectors $\tilde{a}$, $\tilde{b}$ and their first-order partial derivatives with respect to t and the components of U and x are all uniformly continuous and bounded as functions of (t, U, x, ε) if $(t, U, x) \in R$ and ε is small. Let

$$\tilde{A}_0(t, V, y) = \tilde{A}(t, V, y, 0),$$

$$\tilde{B}_0(t, V, y) = \tilde{B}(t, V, y, 0),$$

$$\tilde{a}_0(t, V, y) = \tilde{a}(t, U, x, 0),$$

$$\tilde{b}_0(t, V, y) = \tilde{b}(t, U, x, 0).$$

Then the degenerate system corresponding to (5.52) is

$$\tilde{A}_0(t, V, y)\frac{dV}{dt} + \tilde{a}_0(t, V, y) = 0,$$

$$\frac{dy}{dt} = \tilde{B}_0(t, V, y)\frac{dV}{dt} + \tilde{b}_0(t, V, y). \qquad (5.53)$$

Now we define a discontinuous solution of the degenerate system. Let the point $P_1 = (s_1, V_1, y_1)$ be such that $P_1 \in R$, and assume that the real parts of all the characteristic roots of $A_0(s_1, V_1, y_1)$ are positive. Let $(v(t), y(t))$ be a solution of (5.53) such that

$$(v(s_1), y(s_1)) = (v_1, y_1).$$

Assume that there exists $s_2 > s_1$ such that:

1. The solution $(v(t), y(t))$ is defined for $t \in (s_1, s_2)$.
2. $\lim_{t \uparrow s_2}(v(t), y(t))$ exists and is finite. [Denote that limit by (v_2, y_2).] Assume also that $(s_2, v_2, y_2) \in R$.
3. The real parts of the characteristic roots of $A_0(t, v(t), y(t))$ are all positive for $t \in [s_1, s_2)$.
4. The matrix $A_0(s_2, v_2, y_2)$ has 0 as a simple characteristic root, and all the other characteristic roots of $A_0(s_2, v_2, y_2)$ have positive real parts.

By assumption 4, we can assume that the first row and the first column of $A_0(s_2, v_2, y_2)$ are zero. If $m^1 = n + 1$, then $U = (u, w)$ and $V = (v, r)$ where u and v are scalars and w and r are n vectors, and we can write systems (5.52) and (5.53) in the following form:

$$\varepsilon\frac{d^2u}{dt^2} + g(t, u, w, x, \varepsilon)\frac{du}{dt} + \beta(t, u, w, x, \varepsilon)\frac{dw}{dt}$$
$$+ h(t, u, w, x, \varepsilon) = 0,$$

$$\varepsilon\frac{d^2u}{dt^2} + \alpha(t, u, w, x, \varepsilon)\frac{du}{dt} + A(t, u, w, x, \varepsilon)\frac{dw}{dt}$$
$$+ a(t, u, w, x, \varepsilon) = 0, \qquad (5.54)$$

$$\frac{dx}{dt} = f(t, u, w, x, \varepsilon)\frac{du}{dt} + B(t, u, w, x, \varepsilon)\frac{dw}{dt}$$
$$+ b(t, u, w, x, \varepsilon)$$

and

$$g_0(t, v, r, y)\frac{dv}{dt} + \beta_0(t, v, r, y)\frac{dr}{dt} + h_0(t, v, r, y) = 0,$$

$$\alpha_0(t, v, r, y)\frac{dv}{dt} + A_0(t, v, r, y)\frac{dr}{dt} + a_0(t, v, r, y) = 0,$$

$$(5.55)$$

$$\frac{dy}{dt} = f_0(t, v, r, y)\frac{dv}{dt} + B_0(t, v, r, y)\frac{dr}{dt} + b_0(t, v, r, y),$$

where

$$\tilde{A} = \begin{bmatrix} g & \beta \\ \alpha & A \end{bmatrix}, \qquad \tilde{a} = \begin{bmatrix} h \\ a \end{bmatrix}, \qquad \tilde{B} = (f, B), \qquad \tilde{b} = b,$$

where g, h are real scalars, α and a are column n vectors, β is a row n vector, A is an $n \times n$ matrix, f and b are column m vectors, and B is an $m \times n$ matrix. Similar formulas hold for $\tilde{A}_0$, $\tilde{a}_0$, and so on.

Let

$$P_k = (s_k, v_k, r_k, y_k), \qquad k = 1, 2.$$

We assume that at P_2, we have

$$g_0 = 0, \qquad \alpha_0 = 0, \qquad \beta_0 = 0,$$

$$h_0 < 0, \qquad \left(\frac{\partial g_0}{\partial y}\right) f_0 + \frac{\partial f_0}{\partial v} \neq 0.$$

We assume also that if $t < s_2$ and if t is sufficiently close to s_2, then for $(t, v, r, y) = (t, v(t), r(t), y(t))$, where $(v(t), r(t)) = v(t)$, it is true that

$$g_0 - \beta_0 A_0^{-1}\alpha_0 > 0.$$

Let $y = \phi(v)$ be the solution of the system

$$\frac{dy}{dv} = f_0(s_2, v, r_2, y)$$

such that

$$\phi(v_2) = y_2.$$

Let v_3 be the smallest value of $v > v_2$ such that

$$\int_{v_2}^{v} g_0\big(s_2, v, r_2, \phi(v)\big)\, dv = 0$$

and assume that the curve

$$\mathscr{C} = \big\{ (t, v, r, y) \,|\, t = s_2,\ v_2 \le v \le v_3,\ r = r_2,\ y = \phi(v) \big\}$$

is contained in R. Also we assume that on curve $\mathscr{C}$,

$$\alpha_0 = 0$$

and that the real parts of all the characteristic roots of the matrix $A_0 - g_0 I_n$, where I_n is the $n \times n$ identity matrix, are positive.

Now let $y_3 = \phi(v_3)$ and let $P_3 = (s_2, v_3, r_2, y_3)$. Assume that all the characteristic roots of $A_0(t, v, r, y)$ have positive real parts at P_3. Let $(\hat{v}(t), \hat{r}(t), \hat{y}(t))$ be the solution of (5.55) such that

$$\hat{v}(s_2) = v_3,$$

$$\hat{r}(s_2) = r_2,$$

$$\hat{y}(s_2) = y_3.$$

Assume that there exists $s_3 > s_2$ such that:

1. $(\hat{v}(t), \hat{r}(t), \hat{y}(t))$ is defined for $s_2 \le t \le s_3$ and is contained in R.
2. If $\lambda_j(t),\ j = 1, 2, \ldots, n, n+1$, are the characteristic roots of the matrix

$$\tilde{A}_0(t, \hat{v}(t), \hat{r}(t), \hat{y}(t)),$$

then

$$\operatorname{Re} \lambda_j(t) > 0, \qquad j = 1, 2, \ldots, n, n+1 \text{ and } s_2 \le t \le s_3.$$

Then a *discontinuous solution* S_0 of the degenerate system is the union of the curve

$$\big\{ (t, v, r, y) \,|\, s_1 \le t \le s_2,\ v = v(t),\ r = r(t),\ y = y(t) \big\}$$

the curve $\mathscr{C}$, and the curve

$$\big\{ (t, v, r, y) \,|\, s_2 \le t \le s_3,\ v = \hat{v}(t),\ r = r(t),\ y = \hat{y}(t) \big\}.$$

In stating Sibuya's result, we use the following definitions of the norm of a vector and the norm of a matrix. If $c = (c_1, \ldots, c_n)$, then the norm of c is defined to be

$$|c| = \max_j |c_j|,$$

and if A is an $n \times n$ matrix, then the norm of A is defined to be

$$|A| = \max\{|Ac| \,\big|\, |c| = 1\}.$$

Theorem 18. *If δ_1, δ_2, δ_3, and ε are sufficiently small positive numbers, there is a solution $(u(t), w(t), x(t))$ of (5.54) defined on the interval $s_1 \leq t \leq s_3$ for any initial values satisfying the conditions*

$$|u(s_1) - v(s_1)| + |w(s_1) - r(s_1)| \leq \delta_1,$$

$$\left|\frac{du}{dt}(s_1) - \frac{dv}{dt}(s_1)\right| + \left|\frac{dw}{dt}(s_1) - \frac{dr}{dt}(s_1)\right| \leq \frac{\delta_2}{\varepsilon},$$

$$|x(s_1) - y(s_1)| \leq \delta_3,$$

and as δ_1, δ_2, δ_3, and ε approach zero, the curve $\mathscr{S}$ representing the solution $(u(t), w(t), x(t))$ in (t, u, w, x) space approaches the curve S_0, that is,

$$\lim_{\varepsilon \to 0} \max_{p \in S_0} [d(p, \mathscr{S})] = 0.$$

Also for any fixed $\delta > 0$, the terms

$$|u(t) - v(t)| + |w(t) - r(t)| + |x(t) - y(t)|$$

and

$$\left|\frac{du}{dt}(t) - \frac{dv}{dt}(t)\right| + \left|\frac{dw}{dt}(t) - \frac{dr}{dt}(t)\right|$$

converge uniformly to zero on the intervals

$$s_1 \leq r \leq s_2 - \delta$$

and

$$s_1 + \delta \leq t \leq s_2 - \delta,$$

respectively. Moreover, the terms

$$|u(t) - \hat{v}(t)| + |w(t) - \hat{r}(t)|,$$

$$|x(t) - y(t)|,$$

and

$$\left| \frac{du}{dt}(t) - \frac{dv}{dt}(t) \right| + \left| \frac{dw}{dt}(t) - \frac{dr}{dt}(t) \right|$$

converge uniformly to zero on the interval

$$s_2 + \delta \le t \le s_3.$$

5.5 Partial differential equations

All the mathematics described so far in this chapter is concerned with systems of ordinary differential equations and is intended to be applied to some one of the voltage-clamp models described in Chapters 2–4. We have already seen in Chapter 2 that if we are to obtain a mathematical description of an action potential, then we must use a mathematical model that combines the voltage-clamp model with cable theory. Since the resulting model is a system of partial differential equations, which we termed the full Hodgkin–Huxley equations, it would seem reasonable to proceed now to a description of the mathematics of partial differential equations, that is, theory that would be applicable to the full Hodgkin–Huxley equations and analogous models for other physiological models.

Actually, we will not present such theory because it is in the process of being developed. We have already emphasized, in our discussion of ordinary differential equations, that much of the known theory of differential equations has no direct or useful applications in the study of the Hodgkin–Huxley equations and that further mathematical theory needs to be developed to deal with the problems that arise naturally in the study of physiological models, that is, voltage-clamp models. The situation with the full Hodgkin–Huxley equations is similar except that there is not very much theory, applicable or otherwise. The full Hodgkin–Huxley equations are an example of a system of reaction-diffusion equations. Although reaction-diffusion equations are now the subject of extensive study, they received comparatively little attention until about 20 years ago. The extensive work that is now being done on reaction-diffusion equations [see Fife (1978) for a summary of these studies] suggests that, in time, a large body of theory will be developed that will be applicable to such systems as the full

Hodgkin–Huxley equations. However, at present, many of the mathematical problems concerned with the full Hodgkin–Huxley equations remain partially or totally unsolved.

There is a second reason for not giving an account of the mathematical theory that might be applicable to the full Hodgkin–Huxley equations. That reason lies in the fact that, as work on the Hodgkin–Huxley equation progresses, the results may indicate the best direction for work on the full Hodgkin–Huxley equations. Consequently, to some extent, work on the full Hodgkin–Huxley equations should await results concerning the Hodgkin–Huxley equations.

For these reasons, we will restrict ourselves merely to describing briefly (in Chapter 6) some of the work that has been done on versions of the full Hodgkin–Huxley equations.

6

Mathematical analysis of physiological models

6.1   Introduction

We have already described the numerical analysis of the Hodgkin–Huxley equations (in Chapter 3) and other models (in Chapter 4). In view of the high success of much of this numerical analysis, it is natural to ask why any further mathematical analysis is required. That is, when a model of an electrically excitable cell has been obtained, why not answer all mathematical questions that arise by simply carrying out a numerical analysis? The answer to this question lies partly in the nature of the models that we have considered and partly in the kind of results that can be obtained from numerical analysis. As we have seen, the models considered are empirical descriptions and the constants that appear in them are, at best, reasonably good approximations. In fact, for some purposes, it is well to regard these constants as parameters that have various values. Moreover, the very functions that appear in the models are, in some cases, quite tentatively proposed [see Chapter 2 and Hodgkin and Huxley (1952d, p. 510).] In carrying out a numerical analysis, an entirely specific model must be considered. That is, the forms of the functions and the values of the parameters must be completely specified. Moreover, the numerical analysis gives no information about whether solutions of similar models have the same or similar behavior. As a simple example, suppose that the numerical analysis of one model suggests that there is a unique asymptotically stable periodic solution. (Of course, numerical analysis alone cannot be used to *prove* the existence of a periodic solution.) Now suppose that one of the constants or one of the functions in the model is changed slightly. Does the resulting model also have a unique asymptotically stable periodic solution?

It seems reasonable to guess that there is such a periodic solution, but the question cannot actually be answered by the original numerical analysis. Strictly speaking, the only way to answer the question is to carry out a numerical analysis of the new model. This kind of argument suggests the importance of qualitative analysis (which does not depend on exact numerical values) of the solutions of the models studied. However, it should be emphasized that these arguments have been adduced by physiologists [see Jack, Noble, and Tsien (1975, p. 379)]. They are not just arguments put forth by mathematicians seeking employment!

A second drawback to the numerical analysis is that it does not give us any understanding of the in-the-large behavior of the solutions. For example, the numerical analysis of the Hodgkin–Huxley equations shows that approximate solutions display behavior that describe threshold phenomena, but one obtains no understanding of how or why this behavior occurs. However, as we shall see in this chapter, viewing the FitzHugh–Nagumo equations as a singularly perturbed system makes for a much clearer understanding of the how and why of this behavior.

Two very different kinds of mathematical analysis are needed to deal with models of electrically active cells. First, a qualitative analysis is required for the solutions of the models derived from voltage-clamp experiments: the Hodgkin–Huxley equations and the models of other electrically active cells that are obtained from voltage-clamp experiments. For brevity, we shall term them voltage-clamp models. As pointed out earlier, each of these models describes the relationships among the membrane potential and ionic currents, all regarded as functions of time only. Each of these models is a system of nonlinear ordinary differential equations, and the solution of such a system can be interpreted in two ways. First, the solution can be regarded as describing the membrane potential and the ionic currents (as functions of time) at a fixed position on the axon when the axon is functioning naturally. Second, the solution can be regarded as describing the membrane potential and ionic currents that occur in experiments in which the membrane potential and ionic currents have at a given time the same value at all points along the axon. Such experiments include not only the voltage-clamp experiments, but other experiments described in Chapter 2 in which the membrane potential is allowed to vary [see Hodgkin, Huxley, and Katz (1952)]. If the membrane potential and the ionic currents have the same value at all points along the axon, they are said to be uniform. The term "uniform action potential" refers to a solution of the system for which the membrane potential and the ionic currents are uniform and the initial value of the membrane potential exceeds the threshold value. Then a large, but brief, flow of sodium current occurs and is followed by an outward flow of potassium current.

Since all the voltage-clamp models are systems of nonlinear ordinary differential equations, then the qualitative theory of such differential equations that was described in Chapter 5 can be brought to bear on the analysis of the models. As it turns out, the most effective viewpoint is to regard the system of ordinary differential equations as a singularly perturbed system and to study the solutions that are close to the discontinuous solutions (the theories of Levinson and of Mishchenko and Rozov that were described in Chapter 5). Since the decision to adopt this viewpoint is crucially important to the later analysis, we summarize here the mathematical and physiological reasons for this decision. First, numerical computations of values of the functions that appear in the equations suggest strongly that the equations should be regarded as singularly perturbed. We will see this later in the discussion of the Hodgkin–Huxley equations and the Noble equations. Next, experimental results show that the membrane potential V and the Na^+ current undergo rapid smooth changes. Moreover, numerical study of the solutions of the Hodgkin–Huxley equations and the Noble equations show that the solution component $V(t)$ and the solution component $m(t)$, which is a measure of the Na^+ current, display very rapid smooth changes. These smooth rapid changes are characteristic of the solutions of a singularly perturbed system that are close to discontinuous solutions. Third, numerical studies of the Hodgkin–Huxley equations made by Cole, Antosiewicz, and Rabinowitz (1955) show that if I increases above a particular value, a periodic solution appears, that is, bifurcation occurs. However, as will be discussed in Section 6.2.1.3 of this chapter, the bifurcation that occurs does not look like a Hopf bifurcation, but rather like the kind of bifurcation that occurs in singularly perturbed systems. In such bifurcations, the periodic solution that appears is near a discontinuous solution. Finally, it should be pointed out that discontinuous solutions give quick global pictures of the behavior of the general solutions. This is just the kind of information that the physiologist needs to complement numerical studies.

In the analysis of Hodgkin–Huxley equations in Chapter 2, it was shown that in order to obtain a quantitative description of the natural behavior of the axon, it is necessary to combine the Hodgkin–Huxley equations (i.e., the voltage-clamp equations) with cable theory to derive a system of nonlinear partial differential equations. We referred to this system as the full Hodgkin–Huxley equations. A solution of this system describes the membrane potential and the ionic currents as functions of both time and position along the axon. The search for a solution that described an action potential was shown to be a search for a traveling wave solution and this, in turn, becomes the study of a system of nonlinear ordinary differential equations. But the problems that must be dealt with concerning these nonlinear differential equations are very different from those that arise in the study of the voltage-clamp equations. In the last part of this chapter, we will describe briefly the kinds of problems that arise and some of the work that has been done.

It should be pointed out at this stage that, while numerical analysis has shown that the full Hodgkin–Huxley equations are a valid model of the squid axon, it does not follow that combining voltage-clamp equations with cable theory always yields a valid model. In fact, such a technique does not seem to work for models of the Purkinje fiber [see McAllister et al. (1975, pp. 4 and 53)].

Finally, we must point out that this chapter is inevitably fragmentary. The reason for this is the present state of our knowledge of the subject. As we shall see, this knowledge is the sum of efforts in various directions, and many of these efforts have not been brought to full fruition. Indeed, in some cases, it is not clear whether these are efforts in the right direction. Thus, this chapter is doomed to a certain incoherence. The one virtue of our discussion is that it shows clearly that there are many open problems in this subject.

6.2   Models derived from voltage-clamp experiments
6.2.1   Nerve conduction models
6.2.1.1   *Some numerical analysis*

Although we are primarily concerned with qualitative analysis of the voltage-clamp models, we start by summarizing briefly some numerical analysis of the Hodgkin–Huxley equations. Besides being valuable per se, these numerical results are important because they are suggestive of directions for qualitative study. We have already described in Chapter 2 the results of the numerical analysis carried out by Hodgkin and Huxley in their original paper. Here we describe briefly some later numerical analysis that is suggestive of qualitative study.

The first numerical analysis of the Hodgkin–Huxley equations with a modern computer was carried out by Cole et al. (1955); see also FitzHugh and Antosiewicz (1959). They studied (i) constant currents that are "threshold," that is, produce an action potential, (ii) the minimal rate of increase of current required to produce an action potential, and (iii) repetitive firing, that is, spontaneous periodic uniform action potentials in response to a constant current. For a sufficiently large constant current I, their numerical results suggested the existence of a periodic solution, that is, an infinite train of uniform action potentials. (If there is a periodic solution, the amplitude of the current stimulus does not affect the frequency (or period) of the periodic solution. That is, there is no frequency modulation. But many axons, especially in the sensory systems, require this frequency modulation [see Cohen (1976, pp. 101 and 121)]. This shows that the Hodgkin–Huxley model is not suitable for sensory pulse conduction. Dodge (1972) has discussed this point.) Studies of repetitive firing have also been carried out by Cooley, Dodge, and Cohen (1965) and others [for references, see Rinzel (1979)].

If the bath in which the axon is immersed has a calcium concentration lower than the normal value, then the axon sometimes displays spontaneous periodic action potentials. This experimental phenomenon has been studied mathematically by Huxley (1959), who proposed a modification of the Hodgkin–Huxley equations to describe the axon in a reduced calcium concentration. Huxley also carried out a numerical analysis of these modified equations and obtained periodic solutions. [McDonough (1979), has carried out an analytic study of these modified equations.] More recently, Guttman, Lewis, and Rinzel (1980) have shown experimentally that such repetitive firing can be annihilated by a brief pulse of appropriate magnitude and phase. Under another such pulse, repetitive firing resumes. It is worth observing that these researchers state that their experimental work was motivated by theoretical studies. (References for the theoretical studies are given in the paper; see especially page 389.) This is an example of how theoretical studies do have a contribution to make in physiological research.

Now we describe briefly some numerical results that should be supported by rigorous mathematical reasoning. First, Berkinblit et al. (1970) have carried out computer studies of the Hodgkin–Huxley (HH) equations that model certain phenomena observed in physiological experiments, especially, summation of subthreshold responses and periodic omission of pulses. The first of these consists in studying the effect of adding a periodic sequence of square current pulses that have a subthreshold effect (the charge thus added raises the membrane potential, but not up to the threshold). The second consists in studying the effect of adding a periodic sequence of square current pulses that have a super-threshold effect. The numerical results obtained agree with the physiological observations and thus provide additional evidence of the validity of the HH equations. But if the numerical results could be supported by rigorous mathematical reasoning, this evidence would be stronger and also a clearer understanding might be obtained of the processes involved.

Another interesting and important numerical study that which should be analyzed theoretically is the numerical study of the model due to Adelman and FitzHugh (1975). As we have pointed out earlier (in Chapter 2), one of the drawbacks of the HH equations is that they exhibit periodic solutions that do not correspond to physiological behavior. The existence of such periodic solutions is suggested by numerical analysis [Cole et al. (1955)] and is proved for the FitzHugh–Nagumo equations (see Section 6.2.1.2 of this chapter). In Chapter 2, this point was discussed and we saw that this drawback of the HH equations is related to the fact that the HH equations do not describe the phenomenon of accommodation or adaptation. Mathematically, this means that there is no dependent variable that is sufficiently slow changing. In FitzHugh's terminology, there is no variable of Type 4 [FitzHugh (1969, pp. 4 and 32)]. There seems little doubt that a more accurate or more realistic version of the HH equations would require somehow taking adaptation into account.

Adelman and FitzHugh (1975) introduced a modification of the HH equations that is a significant step in this direction. Their modification is based on results of physiological experiments. First, it has been observed experimentally that K_s, the concentration of potassium ions in the periaxonal space between the outside of the membrane and the inside of the Schwann cell sheath, plays a significant role in the electrical activity of the axon [see Frankenhaeuser and Hodgkin (1956)]. The concentration K_s changes continually due to the potassium ion flow through the membrane and diffusion across the Schwann cell layer [see Adelman, Palti, and Senft (1973)]. Hence, the potassium potential V_K, which depends on K_s (remember the Nernst formula), is a function also of time. Also, Adelman et al. (1973) showed how $\bar{g}_K$ and α_n and β_n are affected by the varying of V_K with K_s. Adelman and Palti (1969) showed that dh/dt, the rate of change of sodium inactivation, depends on K_s. These dependencies were taken into account by Adelman and FitzHugh in a modified form of the HH equations. Actually, they set up three modifications of the HH equations: the first modification introduced K_s; the second introduced K_s, new values of $\bar{g}_K$, and new functions for α_n and β_n; the third introduced K_s, new values of $\bar{g}_K$, and new functions for α_n, β_n, α_h, and β_h.

The third modification, called by Adelman and FitzHugh the fully modified HH equations, which includes all the changes, can be described as follows. Let K_i denote the potassium ion concentration in the interior of the axon, K_o the potassium concentration in the external bulk solution, and let P_K^s denote the permeability to potassium of the diffusion barrier between the periaxonal space and the external bulk solution in centimeters per second. The modified equations are

$$\frac{dV}{dt} = \frac{-1}{C}\left\{ I - \bar{g}_{Na}m^3 h(V - V_{Na}) - \bar{g}_K n^4 (V - V_K) \right.$$
$$\left. - \bar{g}_L(V - V_L) \right\},$$

$$\frac{dm}{dt} = (1 - m)\alpha_m(V) - m\beta_m(V),$$

$$\frac{dh}{dt} = (1 - h)\alpha_h(V) - h\beta_h(V),$$

$$\frac{dn}{dt} = (1 - n)\alpha_n(V) - n\beta_n(V),$$

$$\frac{dK_s}{dt} = \frac{1}{\theta}\left[\frac{I_K}{F} - P_K^s(K_s - K_o) \right]$$
$$= \frac{1}{\theta}\left[\frac{g_K n^4 (V - V_K)}{F} - P_K^s(K_s - K_o) \right],$$

where θ is the radial thickness of the phenomenological periaxonal space, $\bar{g}_K = 62.5 \ \Omega^{-1}/\text{cm}^2$, and

$$V_K = \frac{RT}{F}\ln\frac{(K_s)}{K_i}.$$

The numerical analysis of this system of equations gives a more accurate representation of a number of physiological phenomena including a better description of adaptation during long duration constant current stimulation. That is, finite trains and nonrepetitive responses are obtained.

It is clear that this modified HH system is more complicated than the original HH equations and since qualitative analysis of the HH equations is very limited, there is little hope of making a rigorous qualitative analysis of the modified HH equations. There remains, however, the possibility of making a qualitative comparison of the two systems with a view to understanding why different types of solutions occur.

6.2.1.2   *Analysis of a two-dimensional model*

  The use of qualitative analysis. In Chapter 3, we introduced two-dimensional models that are simplifications of the HH equations. As explained there, the purpose of introducing the simplified model is to make possible a qualitative analysis of the solutions and thus obtain a clearer understanding of the processes that are taking place. Since we are so much concerned here with qualitative analysis, it is worthwhile first to illustrate the advantages of qualitative analysis with a simple familiar example: resonance in a spring-mass system. We consider a spring-mass system without damping. If m is the mass and k is the spring constant, the action of the system is described by the differential equation

$$m\ddot{u} + ku = 0.$$

If $w = \sqrt{k/m}$, then elementary techniques show that the general solution of this equation is

$$u(t) = c_1\cos wt + c_2\sin wt,$$

where c_1, c_2 are constants. Now suppose a periodic external force with period equal to the period of the preceding general solution, say $A\cos wt$, where A is a nonzero constant, is applied to the spring-mass system. The system is now described by the equation

$$m\ddot{u} + ku = A\cos wt \tag{6.1}$$

and the general solution is shown by elementary methods to be

$$u(t) = c_1\cos wt + c_2\sin wt + \frac{A}{2mw}t\sin wt. \tag{6.2}$$

Regardless of the initial conditions, the term $(A/2mw)t\sin wt$ appears in the solution. The graph of $(A/2mw)t\sin wt$, shown in Fig. 6.1 shows how the solution becomes unbounded as $t \to \infty$. It is important to observe here that we have not only determined that the solution becomes unbounded but also that we have a clear picture of how the solution oscillates more and more wildly. Since we have an explicit formula for the solution, then for given initial values (which will determine the constants c_1 and c_2), we can find the value of the solution at any fixed time t. (That is, we have complete quantitative information.) But the qualitative information (how the solution oscillates more and more wildly) is more im-

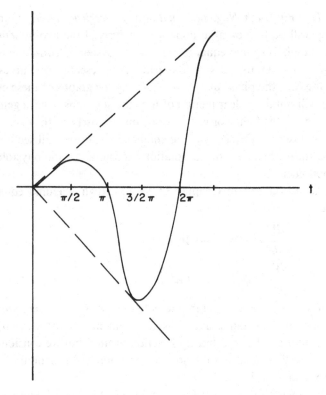

Figure 6.1.

portant in understanding the action of the spring-mass system. Now suppose that instead of solving (6.1) explicitly, we had carried out a numerical analysis. A sufficiently detailed analysis would probably suggest (although, of course, not prove) that the solution oscillates with increasing wildness. But the numerical analysis might not give us as explicit qualitative information. Moreover, we would have little or no understanding of why it occurred.

Since there is no hope of obtaining, for the differential equations that we study, explicit or closed solutions [i.e., solutions like solution (6.2) of equation (6.1)] we try to search directly for qualitative information. The primary purpose of this chapter is to show how the techniques described in Chapter 5 can be used for such a search.

The FitzHugh–Nagumo equations: *Numerical results*. Our first step will be to look at a numerical analysis of the two-dimensional FitzHugh–Nagumo equations. Since the system is two-dimensional and autonomous, its solutions can be represented by curves in a plane (i.e., the phase plane). By studying the graphs of these curves, we will obtain a clear picture of how the solutions behave generally and how this behavior corresponds quite closely to the experimentally observed physiological phenomena. Thus, we will see how the qualitative behavior of the solutions accounts for the physiological phenomena.

As introduced in Chapter 3, the FitzHugh–Nagumo equations are

$$\frac{dV}{dt} = V - \frac{V^3}{3} - W + I,$$

$$\frac{dW}{dt} = \phi(V + a - bW), \tag{6.3}$$

where V denotes the membrane potential, W is a recovery variable, a, b, ϕ are constants, and I is the total membrane current. The current I can be an arbitrary function of time, but we consider first the case that describes the normal condition and functioning of the nerve, that is, $I = 0$.

Following FitzHugh (1969) we take $a = 0.7$, $b = 0.8$, and $\phi = 0.08$ in (6.3). Then by numerical methods, approximate solutions of (6.3) can be computed. The phase plane portrait of the computed solutions is shown in Fig. 6.2 [which is Fig. 3.2 in FitzHugh (1969)]. The dashed curves are the isoclines, that is, the curves

$$\frac{dV}{dt} = V - \frac{V^3}{3} - W = 0$$

and

$$\frac{dV}{dt} = V + a - bW = 0.$$

The critical points or singular points or equilibrium points are the points of intersection of the isoclines, that is, the points (V, W) such that

$$V - \frac{V^3}{3} - W = 0$$

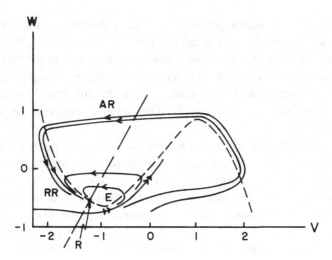

Figure 6.2.

and

$$V + a - bW = 0.$$

In this case, there is just one critical point R, which has coordinates

$$(V_R, W_R) = (-1.1994, -0.6243).$$

The solution curves, with arrowheads to indicate the direction of increasing t, are given by solid curves. The curve with double arrowheads, called the threshold separatrix, separates solution curves that represent action potentials from those that do not represent action potentials. Since the results shown in Fig. 6.2 summarize numerical analysis, they are not all sharply defined. In particular, the threshold separatrix is not well defined. As shown in Fig. 6.2, all the solution curves approach critical point R as time increases. Now suppose that we consider a solution that has the initial value $(-1.1994 + V_0, -0.6243)$, that is, a solution that describes what takes place if the initial value of V is $V_R + V_0$ and the initial value of W is W_R. If V_0 is positive but small enough so that the point $(V_R + V_0, W_R)$ is on the left of the threshold separatrix, then as Fig. 6.2 shows, the solution curve moves almost directly toward the critical point R. This is a solution that does not describe an action potential, that is, it describes a passive local reaction. On the other hand if the initial point $(V_R + V_0, W_R)$ is on the right of the threshold separatrix, then, as the graph of the solution curve shows, $V(t)$ increases rapidly as time passes; then $W(t)$ increases, $V(t)$ decreases, and the solution curve moves toward the critical point R. This solution describes an action potential.

The behavior of these solutions shows how the phase plane can be made into a kind of map of different physiological states. For example, the region labeled AR (for absolutely refractory) consists of points (V, W) such that if a shock is applied so that V is increased to a larger value $\overline{V}$, then no action potential will occur no matter how large $\overline{V}$ is. That is, each point in the region corresponds to a state of the axon in which the axon is absolutely refractory. Similarly, the points in the region labeled RR (for relatively refractory) correspond to states of the axon in which it is relatively refractory. Also the points in the region labeled E (for enhanced) correspond to states of the axon in which it is enhanced, that is, will respond with an action potential even though the shock applied is not large enough to raise the membrane potential from V_R to the usual threshold level. The regions labeled RR, AR, and E in Fig. 6.2 are not indicated very clearly; their boundaries are simply suggested. The reason for this lack of definiteness is that we are summarizing results of numerical analysis and, hence, the results are not sharply delineated; rather the results suggest the existence of such regions.

The relationships among these solutions of (6.3) that we have seen give us an "in-the-large" understanding of the various actions of the membrane potential that could not be achieved from numerical analysis of the HH equations. Of course, we have paid a price for this qualitative understanding: Much of the realism of the description given by the HH equations has been sacrificed. The activation and inactivation variables that describe quantitatively the currents of sodium and potassium ions in the HH equations have been replaced by a recovery variable W that has no specific physiological meaning. Nevertheless, the derivation and analysis of the FitzHugh–Nagumo equations represents a large forward step in the quantitative analysis of nerve conduction.

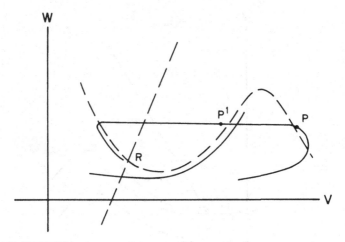

Figure 6.3.

Numerical analysis of the FitzHugh–Nagumo model can also be used [FitzHugh (1961)] to explain a number of other properties of the nerve membrane: for example, abolition of an action potential by anodal shock, and anodal break excitation.

Suppose that an action potential occurs and is described by the solid line in Fig. 6.3. If an anodal shock is applied at the instant that the action potential is described by the point P, then the phase point is displaced horizontally to the left of P to a new point P', and if this displacement is large enough so that P' is on the left of the threshold separatrix (as sketched), then the phase point continues on the solid line back to the equilibrium point R. Thus, the action potential is abolished by the anodal shock. This phenomenon has been observed experimentally: See Blair and Erlanger (1935) and Tasaki (1955).

Next suppose that a constant membrane current I, where $I < 0$, is applied for an interval of time of length, say, T. Then the V isocline becomes

$$V - \frac{V^3}{3} - W + I = 0$$

and its graph is obtained simply by lowering the V isocline in Fig. 6.2, by $|I|$ units. Then the new equilibrium point R' is below R (see

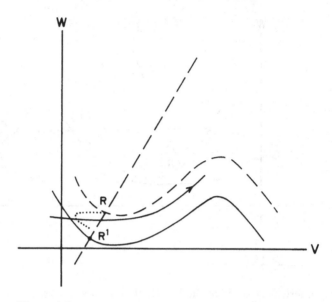

Figure 6.4.

Fig. 6.4). If the phase point starts at R when I is applied, it will move toward R' along the dotted line in Fig. 6.4. Now if $|I|$ and T are large enough, then at the end of the time interval of length T, the phase point will be below the threshold separatrix and so when the membrane current I is made zero again, an action potential will occur. This is the phenomenon of anodal break excitation that was discussed earlier (Chapter 2). In Chapter 2, we stated that appropriate numerical solutions of the Hodgkin–Huxley equations describe fairly accurately anodal break excitation. Also we presented an explanation in words of why anodal break excitation occurs. It is interesting to compare the explanations in Chapter 2 with the preceding explanation, which is quantitative, not just an explanation in words; also it relates the phenomenon of anodal break excitation to other phenomena.

FitzHugh also made a motion picture entitled "Impulse propagation in a nerve fiber," which is based on numerical study of equation (6.3). This motion picture, which is very helpful in understanding how an action potential occurs, is available on loan from the National Medical Audiovisual Center (Annex), Station K, Atlanta, Georgia 30333.

The FitzHugh–Nagumo equations: *Singular perturbation analysis.*
FitzHugh's work is based entirely on numerical analysis, and our
next step is to back up that numerical analysis with rigorous
mathematical theory. We will use singular perturbation theory to
support the numerical analysis and also to give a clearer understanding of the behavior of the solutions. Further examination of
Fig. 6.2 indicates a very important property of the FitzHugh–
Nagumo equations: The solutions display behavior that is typical
of solutions of a singularly perturbed system. When considered
from this viewpoint, the behavior of the solutions of the
FitzHugh–Nagumo equations is easier to understand. In particular,
the origin of the threshold separatrix becomes clear. In order to see
this, we show first how to change variables so that the
FitzHugh–Nagumo equations become a singularly perturbed system.

Let $\tau = t\phi$. Then (6.3) becomes

$$\frac{dV}{dt} = \frac{dV}{d\tau}\frac{d\tau}{dt} = \frac{dV}{d\tau}\phi = V - \frac{V^3}{3} - W,$$

$$\frac{dW}{dt} = \frac{dW}{d\tau}\frac{d\tau}{dt} = \frac{dW}{d\tau}\phi = \phi(V + a - bW)$$

or

$$\frac{dV}{d\tau} = \frac{1}{\phi}\left(V - \frac{V^3}{3} - W\right),$$

$$\frac{dW}{d\tau} = V + a - bW. \tag{6.3'}$$

Since $\phi = 0.08$, the (6.3') has the conventional form of a singularly
perturbed system of ordinary differential equations, that is, a
system of the form

$$\frac{dx}{dt} = \frac{1}{\varepsilon}f(x, y),$$

$$\frac{dy}{dt} = g(x, y),$$

where ε is small. Hence, we may apply the theorems described in
Section 5.2.4 of Chapter 5.

For simplicity of notation we write t in place of τ in (6.3'). Then since $a = 0.7$ and $b = 0.8$, (6.3') becomes

$$\frac{dV}{dt} = \frac{1}{\varepsilon}\left(V - \frac{V^3}{3} - W\right),$$

$$\frac{dW}{dt} = V + 0.7 - 0.8W, \qquad (6.3'')$$

where we have replaced $\phi = 0.08$ by the small positive parameter ε. The graph associated with (6.3'') is indicated in Fig. 6.5. Except very close to the curve

$$V - \frac{V^3}{3} - W = 0,$$

the directions of the solution curves are nearly horizontal and they go to the left or the right (according as $V - V^3/3 - W \lessgtr 0$) as indicated by arrows in Fig. 6.5. Also, on the curve

$$V - \frac{V^3}{3} - W = 0,$$

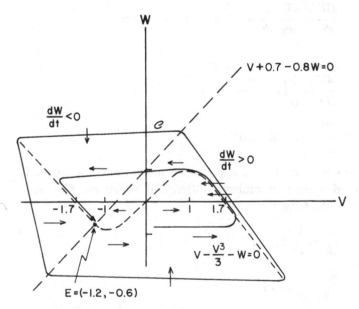

Figure 6.5.

dW/dt is positive on the right of the equilibrium point $E = (-1.2, -0.6)$ and negative on the left of E, as indicated in Fig. 6.5.

Notice that if we take a sufficiently large contour $\mathscr{C}$ (sketched in Fig. 6.5), every solution that intersects $\mathscr{C}$ crosses it and goes into the interior of $\mathscr{C}$. Thus all solutions ultimately enter a bounded set, the interior of $\mathscr{C}$.

Now we show that in a certain approximate sense the solutions of (6.3″) approach the equilibrium point E. First, the point E is an asymptotically stable equilibrium point of (6.3″). This is because the linear part of the right-hand side in a neighborhood of E is

$$\begin{bmatrix} \dfrac{1}{\varepsilon}\left[1 - (-1.2)^2\right] & -\dfrac{1}{\varepsilon} \\ 1 & -0.8 \end{bmatrix} = \begin{bmatrix} \dfrac{1}{\varepsilon}[-0.44] & -\dfrac{1}{\varepsilon} \\ 1 & -0.8 \end{bmatrix}.$$

Since the trace of this matrix is negative and the determinant is positive, it follows that the eigenvalues of the matrix have negative real parts. Hence, by a standard theorem, equilibrium point E is asymptotically stable. That is, if ε is fixed, then $\varepsilon_1 > 0$ implies there exists $\delta > 0$ such that if $(V(t), W(t))$ is a solution of (6.3″) and if $(V(t_0), W(t_0)) \in N_\delta(E)$, then for all $t > t_0$, $(V(t), W(t))$ is defined and

$$\left(V(t), W(t)\right) \in N_{\varepsilon_1}(E).$$

Actually we can say more. The δ clearly depends on ε_1, but δ is independent of ε. To see this, we simply take $\varepsilon = 1$, conclude that E is asymptotically stable, and then obtain the corresponding Lyapunov function $\mathscr{V}$. But it is easy to see, because of the simple role that ε plays in (6.3″), that Lyapunov function $\mathscr{V}$ "works" for all values of ε. The neighborhood $N_\delta(E)$ can be defined in terms of an inequality of the form

$$\mathscr{V} < \varepsilon_2,$$

where $\varepsilon_2 > 0$.

Next, we observe that there exists a positive number T such that if S is a discontinuous solution defined on the interval $[\alpha, \alpha + T]$ with the initial point of S contained in the set Int($\mathscr{C}$), then the point of S corresponding to $\alpha + T$ is contained in $N_{\delta/2}(E)$. That is, all discontinuous solutions that have initial point $p \in$ Int($\mathscr{C}$) arrive in $N_{\delta/2}(E)$ after the time interval T.

Finally by Theorem 15R in Chapter 5, if $q \in \text{Int}(\mathscr{C})$, then if ε is sufficiently small, there is a solution $(V(t), W(t))$ of (6.3″) that is defined for $[\alpha, \alpha + T]$, which is such that $(V(\alpha), W(\alpha)) = q$ and is such that

$$(V(\alpha + T), W(\alpha + T)) \in N_\delta(E).$$

Hence, the solution is defined for all $t \geq \alpha$ and for $t \geq \alpha + T$, the solution remains in the neighborhood $N_{\varepsilon_1}(E)$.

This shows that the equilibrium point has a kind of global asymptotic stability property. Of course, this conclusion is valid only if ε is sufficiently small, but since the FitzHugh–Nagumo equations are intended to give only a qualitative description, we may choose a value of ε for which the conclusion is valid; that is, there is no physiological evidence that limits us to the value

$$\varepsilon = \phi = 0.08.$$

(On the other hand if we try to use Levinson's theorems or the Mishchenko–Rozov theorems in analyzing models that are derived from quantitative physiological data, that is, models like the HH equations, then we do not have such freedom in the choice of ε.)

Now suppose we describe the situation in which a constant current I, where $I > 0$, is impressed on the membrane. Then the system (6.3) is

$$\frac{dV}{dt} = V - \frac{V^3}{3} - W + I,$$

$$\frac{dW}{dt} = \phi(V + a - bW)$$

and the system (6.3″) becomes

$$\frac{dV}{dt} = \frac{1}{\varepsilon}\left(V - \frac{V^3}{3} - W + I\right),$$

$$\frac{dW}{dt} = V + 0.7 - 0.8W,$$

where a, b are taken as 0.7 and 0.8, respectively. The graph associated with this system is indicated in Fig. 6.6. Roughly speaking, it is obtained by "lifting" the cubic curve a vertical distance

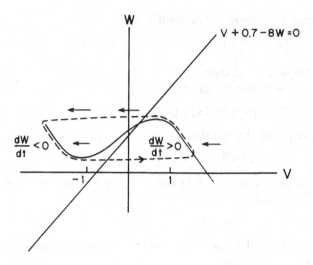

Figure 6.6.

equal to *I*. If the cubic curve is lifted enough so that the line

$$V + 0.7 - 0.8W = 0$$

intersects the cubic curve between its minimum and maximum points (as indicated in Fig. 6.6), then there is a discontinuous solution of the degenerate system that is a closed curve (sketched as dashed curve).

Now we rewrite (6.3″) in the notation of the Mishchenko–Rozov theorems described in Chapter 5, and it becomes

$$\frac{dx}{dt} = \frac{1}{\varepsilon}\left(x - \frac{x^3}{3} - y + I\right),$$

$$\frac{dy}{dt} = x + 0.7 - 0.8y.$$

That is,

$$f(x, y) = x - \frac{x^3}{3} - y + I,$$

$$g(x, y) = x + 0.7 - 0.8y.$$

Hypothesis 1 of the Mishchenko–Rozov theorem (Theorem 14 of

Chapter 5) is obviously satisfied. Since

$$[f_x]^2 + [f_y]^2 = (1 - x^2)^2 + (-1)^2 \geq 1 > 0,$$

hypothesis 2 is satisfied.

The nonregular points of

$$\Gamma = \{(x, y) | f(x, y) = 0\}$$

are the points that satisfy

$$f_x(x, y) = 1 - x^2 = 0,$$

that is, the points $(1, \frac{2}{3} + I)$ and $(-1, \frac{4}{3} + I)$. Thus the nonregular points are isolated. Since

$$f_{xx}(x, y) = -2x,$$

then each of the nonregular points is nondegenerate. Thus, hypothesis 3 is satisfied.

The stable parts of Γ are the points (x, y) with $x \leq -1$ and the points with $x \geq 1$. As Fig. 6.6 indicates, the line

$$g(x, y) = 0$$

does not intersect the stable part of Γ or the nonregular points. Thus, hypothesis 4 is satisfied.

The two nonregular points have different ordinates. Hence, hypothesis 5 is satisfied. As already indicated, there is a closed discontinuous solution, that is, hypothesis 6 is satisfied. Therefore, we may apply the Mishchenko–Rozov theorem (Theorem 14, Chapter 5) and conclude that there is a unique stable periodic solution of the singularly perturbed system for each sufficiently small positive ε.

Finally, let us suppose that I is large enough so that the cubic curve is lifted so much that the line

$$V + 0.7 - 0.8W = 0$$

intersects the cubic curve on the right of the minimum and the maximum points (as indicated in Fig. 6.7). Then as the discontinuous solutions (sketched as dashed curves) indicate, the equilibrium point E is globally asymptotically stable.

As a second example of singular perturbation analysis, we consider the mathematical description of the results of experiments in which the nerve fiber is immersed in a solution bath of reduced calcium concentration. It has long been known that if the calcium

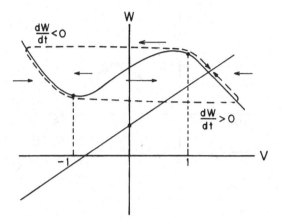

Figure 6.7.

concentration is reduced, the axon may display spontaneous oscillations or repetitive activity. A quantitative study of this activity was made by Frankenhaeuser and Hodgkin (1957) and a mathematical analysis of their results was carried out by Huxley (1959).

Frankenhaeuser and Hodgkin found that the main results could be "summarized by saying that the effects of a fivefold reduction of calcium on the system controlling Na and K permeability are similar to those of a depolarization of 10–15 mV." Huxley proposed describing the corresponding change in the mathematical model, that is, the Hodgkin–Huxley equations, by replacing V in the equations for dm/dt, dh/dt, and dn/dt by $V + \Delta V$, where

$$\Delta V = k \ln \frac{[Ca]}{[Ca]_n},$$

where $k = 9.32$ and $[Ca]$ is the calcium concentration and $[Ca]_n$ is the normal calcium concentration. (Thus if the calcium concentration is lower than $[Ca]_n$, then ΔV is negative.) Hence, the effect in the FitzHugh–Nagumo equations [written in the form (6.3″)] would be to change them to

$$\frac{dV}{dt} = \frac{1}{\varepsilon}\left[V - \frac{V^3}{3} - W\right],$$

$$\frac{dW}{dt} = (V - \tilde{R}) + 0.7 - 0.8W,$$

(6.4)

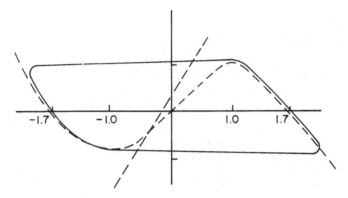

Figure 6.8.

where $\tilde{R} = -\Delta V$. (Remember that the variable W corresponds to the variables m, h, n governing the flow of Na and K ions.)

The graph associated with system (6.4) differs from the graph associated with the original system (6.3″), that is, Fig. 6.3, in that the diagonal line described by

$$V + 0.7 - 0.8W = 0$$

is replaced by the line described by

$$V - \tilde{R} + 0.7 - 0.8W = 0.$$

That is, the diagonal line is moved to the right by distance $\tilde{R}$. Thus if,

$$-1 < R < 2,$$

the graph is as in Fig. 6.8. Thus, a relaxation oscillation of the form sketched in Fig. 6.8 occurs. Finally, if $R > 2$, the graph is as in Fig. 6.9. Thus, as indicated in that graph, the equilibrium point p is globally asymptotically stable.

On the other hand, if R is negative, that is, if the calcium concentration [Ca] is increased, then the diagonal line is moved to the left and the graph has the form shown in Fig. 6.10. Thus, the equilibrium point is globally asymptotically stable.

A stochastic model of nerve conduction: *The model of Lecar and Nossal*. Next we describe another two-dimensional model that is also a modification of the HH equations, but that is a study in a very different direction – a direction that should be pursued further. In order to motivate the introduction of this model, we point out

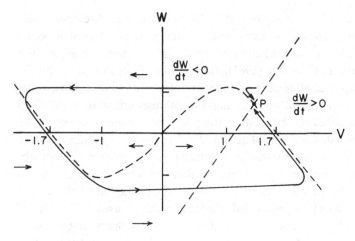

Figure 6.9.

two drawbacks of the HH equations. First, there is a significant difference between what is observed experimentally when a stimulus just equal to the threshold value is applied to an axon and the prediction made by solutions of the HH equations of what will occur. The functions that appear in the HH equations are "well-behaved" functions and hence the basic existence and uniqueness theorems for solutions of systems of differential equations are applicable. Because of the uniqueness of the solution, the HH equations always make the same prediction concerning the re-

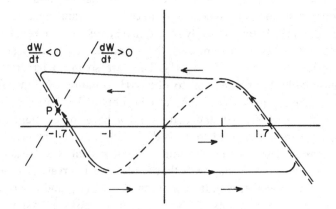

Figure 6.10.

sponse of the axon to a stimulus equal to the threshold value. However, the experimental result is that an action potential occurs for only a certain fraction of stimuli of threshold value [see Ten Hoopen and Verveen (1963)]. Second, the all-or-none law, which is an empirical statement or summary of experimental results, states that the axon either does not fire or responds with a full-sized action potential, but computation of approximate solutions of the HH equations shows that stimuli of various magnitudes produce intermediate responses. Such intermediate responses are not observed experimentally. These conflicts between experimental observation and theoretical prediction can be explained if we take into account an unjustified assumption that is used in the derivation of the HH equations: the assumption that the resting potential is a constant. Actually, experimental results show that the resting potential is subject to small random fluctuations or noise [see Verveen and Derksen (1968)]. The processes by which the resting potential is maintained are not very well understood, but theories of how these processes occur suggest that the processes are of a discrete nature and, thus, it is not surprising that such noise should exist.

These remarks suggest that the HH equations might yield more realistic or accurate predictions if a random or stochastic term were included in the equations. The model that we will describe is a first step in this direction. It was proposed and studied by Lecar and Nossal (1971). Their work included analysis of both the HH equations and the model introduced by Frankenhaeuser and Huxley (1964) for the study of the myelinated axon (see Chap. 4, Sec. 4.2). In both these models, the quantities $\tau_h(V)$ and $\tau_n(V)$ are considerably larger than $\tau_m(V)$. They are so much larger that a reasonable approximation to the HH equations or the Frankenhaeuser–Huxley equations can be obtained by assuming that $dh/dt = 0$, $dn/dt = 0$, and that h and n have the constant values $h_\infty(V_0)$ and $n_\infty(V_0)$, where V_0 is the resting value of the membrane potential. The resulting two-dimensional system is a "reasonable approximation" in the sense that for limited intervals of times its solutions, together with the conditions $h = h_\infty(V_0)$ and $n = n_\infty(V_0)$, are near the solutions of the full four-dimensional system. The reason that the approximation works fairly well is that V and m

vary much more rapidly than h and n. This kind of approximation was introduced by FitzHugh (1961) and was termed the method of reduction.

First we describe this method very briefly. A more detailed description is given in Section 6.2.1.3. Following any stimulus of the axon, the quantities V, m, h, n ultimately return to their resting values. Thus, we expect that the HH equations have a unique globally asymptotically stable equilibrium point. The reduced system is:

$$\frac{dV}{dt} = \frac{1}{c}\left\{ I - \bar{g}_{Na}m^3\bar{h}(V - V_{Na}) - \bar{g}_K\bar{n}^4(V - V_K) - \bar{g}_L(V - V_L) \right\},$$

$$\frac{dm}{dt} = \frac{m_\infty(V) - m}{\tau_m(V)},$$

where

$$\bar{h} = h_\infty(V_0),$$
$$\bar{n} = n_\infty(V_0),$$

and it can be shown [see FitzHugh (1969)] that the reduced system has three equilibrium points: two asymptotically stable equilibrium points and a saddle point. See Fig. 6.11, where the points A, B, C are the equilibrium points. The point A corresponds to the resting state of the system; B is the saddle point. The dashed line is the threshold separatrix. Solutions that pass through a point on the right of the dashed line approach the equilibrium point C as

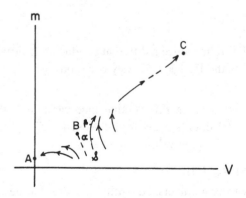

Figure 6.11.

$t \to \infty$, and solutions that pass through a point on the left of the dashed line approach the equilibrium point A as $t \to \infty$. Thus the saddle point and the threshold separatrix provide a model of a system with threshold behavior.

The solutions do not approach a unique globally asymptotically stable equilibrium point as would be expected for the HH equations. The reason for this is that h and n largely describe or determine the later behavior of the solutions of the HH equations. Since h and n are "suppressed" in the reduced system, we can no longer expect the behavior that occurs with the HH equations.

Now the analysis carried out by Lecar and Nossal is the following. A stochastic or random term is added to each of the equations in the reduced system. The resulting system is linearized in a neighborhood of the saddle point B and the linear system thus obtained is studied. The first step in describing these results is to rewrite the reduced system in a different and more convenient notation. The following notation is introduced:

$$\lambda_0 = C^{-1}\left[\bar{g}_K(\bar{n})^4 + \bar{g}_L\right],$$

$$\lambda_1 = C^{-1}\left[\bar{g}_{Na}\bar{h}\right],$$

$$\lambda(V) = \frac{1}{\tau_m(V)},$$

$$V_0 = \frac{\bar{g}_K(\bar{n})^4 V_K + g_L V_L}{\bar{g}_K(\bar{n})^4 + g_L},$$

$$V_1 = V_{Na},$$

$$J = C^{-1}I.$$

Lecar and Nossal also introduce a different conductance variable σ that, for the case of the Hodgkin–Huxley equations, is

$$\sigma = m^3.$$

Their reason for introducing this conductance variable is that a study of experimental data [see Lecar and Nossal (1971, p. 1053)] suggests that the steady state value of σ, that is,

$$\sigma_\infty(V) = \left[m_\infty(V)\right]^3,$$

seems more likely to be a universal quantity (i.e., one that describes data from a wide class of experiments) than the steady state value of m.

With these notations, the reduced system becomes

$$\frac{dV}{dt} = J - \gamma_0(V - V_0) - \gamma_1\sigma(V - V_1),$$
$$\frac{d\sigma}{dt} = \phi(V, \sigma),$$

(6.5)

where, in the case of the HH equations,

$$\phi(V, \sigma) = 3\lambda(V)\sigma^{2/3}\{[\sigma(V)]^{1/3} - \sigma^{1/3}\}.$$

[Lecar and Nossal emphasize that much of the analysis that follows can be carried out without reference to the particular form of $\phi(V, \sigma)$.]

For the Frankenhaeuser–Huxley model, a similar reduced system can be derived. It has the form

$$\frac{dV}{dt} = \frac{1}{C}\{I - I_i\},$$
$$\frac{d\sigma}{dt} = \phi(V, \sigma),$$

(6.6)

where

$$I_i = P_K(\bar{n})^2 f(V - V_K) + \bar{g}_L(V - V_L)$$
$$\qquad + P_{Na}m^2\bar{h}f(V - V_{Na}) + P_p\bar{p}^2f(V - V_p),$$
$$\bar{n} = n_\infty(V_0),$$
$$\bar{h} = h_\infty(V_0),$$
$$\bar{p} = p_\infty(V_0),$$
$$\sigma = m^2,$$

and

$$\phi(V, \sigma) = \frac{2\sigma^{1/2}}{\tau_m(V)}\{[\sigma_\infty(V)]^{1/2} - \sigma^{1/2}\}.$$

Now if noise is taken consideration, then random or stochastic terms must be added to each equation in the reduced system (6.5) or the system (6.6). Of course, once this is done, we are no longer concerned with the problem of solving a system of differential equations, but instead a system of stochastic differential equations. Instead of looking for a deterministic solution $(V(t), \sigma(t))$ of (6.5), we must seek a solution of the system of stochastic differential equations, $(\mathscr{V}(t), \Sigma(t))$, where for each fixed value of t, say t_0, the expressions $\mathscr{V}(t_0), \Sigma(t_0)$ represent random variables. That is, $(\mathscr{V}(t), \Sigma(t))$ is a time-varying probability distribution for the position of the system in the $(V\sigma)$ plane.

Adding random terms to the system (6.5), we obtain

$$\frac{dV}{dt} = J - \gamma_0(V - V_0) - \gamma, \sigma(V - V_1) + F_V(\langle\sigma\rangle, t), \quad (6.7)$$

$$\frac{d\sigma}{dt} = \phi(V, \sigma) + F_\sigma(\langle V\rangle, t), \quad (6.8)$$

where F_V and F_σ are random functions of time (Langevin forces); F_V is assumed to depend on $\langle\sigma\rangle$, the mean value of σ, and F_σ is assumed to depend on $\langle V\rangle$, the mean value of V.

Now (6.7) and (6.8) are expanded in a neighborhood of the equilibrium point $B = (V_B, \sigma_B)$. Using the dimensionless variables,

$$\varepsilon = \frac{V - V_B}{V_1},$$

$$\mu = \sigma - \sigma_B,$$

the expanded forms of (6.7) and (6.8) become

$$\frac{d\varepsilon}{dt} = A_{11}\varepsilon + A_{12}\mu + \cdots + \frac{1}{V_1}F_V(\sigma_B, t) + \cdots, \quad (6.9)$$

$$\frac{d\mu}{dt} = A_{21}\varepsilon + A_{22}\mu + \cdots + F_\sigma(V_B, t) + \cdots, \quad (6.10)$$

where

$$A_{11} = -(\gamma_1\sigma_B + \gamma_0),$$

$$A_{12} = \gamma\frac{1}{1 - V_1}(V_1 - V_B),$$

$$A_{21} = \lambda(V_B)V_1\left[\frac{\partial\sigma_\infty(V)}{\partial V}\right]_{V=V_B},$$

$$A_{22} = -\lambda(V_B).$$

The linearized version of (6.9) and (6.10) is

$$\frac{d\varepsilon}{dt} = A_{11}\varepsilon + A_{12}\mu + \frac{1}{V_1}F_V(\sigma_B, t), \quad (6.11)$$

$$\frac{d\mu}{dt} = A_{21}\varepsilon + A_{11}\mu + F_\sigma(V_B, t), \quad (6.12)$$

and it is this last system that Lecar and Nossal analyze in detail. Thus their conclusions are based on the following assumptions:

1. The system (6.5) is a good approximation for the HH equations; and similarly the system (6.6) is a good approximation for the Frankenhaeuser–Huxley equations (as indicated earlier, this assumption is valid if the interval is small).

2. The linear system (6.11) and (6.12) is a good approximation for the system (6.9) and (6.10).

(Since both eigenvalues of the matrix

$$\begin{bmatrix} A_{11} & A_{12} \\ A_{21} & A_{22} \end{bmatrix},$$

have nonzero real parts, then a sufficient condition that assumption 2 be valid is that ε and μ be small enough.)

It is also assumed that $F_V(\sigma_B, t)$ and $F_\sigma(V_B, t)$ each represents the result of many small independent random events and, hence, for each t is a Gaussian or normal random variable. (Regarding F_V and F_σ, for each t, as normal random variables is suggested by the central limit theorem.)

On the basis of these assumptions, Lecar and Nossal were able to obtain several very interesting results. For example, they calculated the probability of firing (occurrence of an action potential) for a myelinated axon given the initial value of V. Their results agree with experimental results [see Lecar and Nossal (1971, p. 1058)]. Another result is the distribution of latency values. Latency is the physiological term for the time interval from the beginning of the stimulus to the appearance of an action potential (measured, say, at the time at which it reaches half its full height). Experimental results usually suggest that the latency is bounded, but in the saddle point model of threshold behavior the latency is unbounded. The reason for this is that on an orbit that passes very close to the saddle point, such as the orbit $\mathscr{S}$ in Fig. 6.11 the speed (i.e., $[(dV/dt)^2 + (dm/dt)^2]^{1/2}$) is very small because at the saddle point,

$$\frac{dV}{dt} = \frac{dm}{dt} = 0.$$

Hence, the time required to pass on $\mathscr{S}$ from point α to point β is very large; but Lecar and Nossal's results show that the probability of large latencies is very small. Thus the model is more realistic. In other words, the addition of random terms in the differential equations makes the saddle point a more accurate or realistic model of threshold behavior.

The work of Lecar and Nossal is the only work in the direction of including a stochastic term in the mathematical model, and it raises many questions, but it slows clearly a direction of study that should be pursued more extensively.

6.2.1.3 *Analysis of the Hodgkin–Huxley equations*

Boundedness of solutions. The analysis will be described in terms of the following version of the Hodgkin–Huxley equations given by FitzHugh (1969):

$$\frac{dV}{dt} = -\frac{1}{C}\left\{ \bar{g}_{Na}m^3h(V - V_{Na}) + \bar{g}_K n^4(V - V_K) + \bar{g}_L(V - V_L)\right\},$$

$$\frac{dm}{dt} = \frac{m_\infty(V) - m}{\tau_m(V)},$$

$$\frac{dh}{dt} = \frac{h_\infty(V) - h}{\tau_h(V)},$$

$$\frac{dn}{dt} = \frac{n_\infty(V) - n}{\tau_n(V)}.$$

We notice first that if $m = 0$, dm/dt is positive and if $m = 1$, dm/dt is negative, and hence every physiologically significant solution of the Hodgkin–Huxley equations stays in the set

$$\{(V, m, h, n)|0 \leq m \leq 1\}.$$

Similar arguments hold for h and n and we conclude that all solutions of interest stay in the set

$$\{(v, m, h, n)|m \in [0,1], \ h \in [0,1], \ n \in [0,1]\}.$$

Now let

$$\overline{V} = \max[V_{Na}, V_K, V_L],$$

$$\underline{V} = \min[V_{Na}, V_K, V_L].$$

If $V > \overline{V}$, then $dV/dt < 0$ and if $V < \underline{V}$, $dV/dt > 0$. Thus, if r is a positive number, it follows that all the physiologically significant solutions stay in the closed bounded set

$$\left\{ (V, m, h, n) | \underline{V} - r \leq V \leq \overline{V} + r, 0 \leq \begin{matrix} m \\ h \\ n \end{matrix} \leq 1 \right\}.$$

Reduction method of FitzHugh. In view of the ease with which the desired boundedness condition on the solutions is established, it is rather a surprise to find that further analysis of the HH equation is extremely difficult. We will describe a couple of approaches that have been used to obtain some qualitative understanding of the solutions. Both of these approaches depend upon the fact that $\tau_m(V)$ is about one-tenth the size of $\tau_h(V)$ and $\tau_n(V)$. The first approach, the reduction method of FitzHugh, consists in first approximating the HH equations by taking $dh/dt = 0$ and $dn/dt = 0$, that is, by taking τ_h and τ_n to be infinite and letting h and n be equal, respectively, to $h_\infty(0)$ and $n_\infty(0)$. Then the HH equations are reduced to the two-dimensional system

$$\frac{dV}{dt} = \frac{1}{C} \Big\{ \bar{g}_{Na} m^3 [h_\infty(0)](V - V_{Na})$$

$$+ \bar{g}_K [n_\infty(0)]^4 (V - V_K) + \bar{g}_L(V - V_L) \Big\},$$

$$\frac{dm}{dt} = \frac{m_\infty(V) - m}{\tau_m(V)}.$$

Numerical analysis [see FitzHugh (1969)] shows that this system has three equilibrium points. A sketch of these equilibrium points and some nearby orbits is given in Fig. 6.12. [The sketch is not numerically accurate. For a more accurate representation, see FitzHugh (1969).] Equilibrium points A and C are stable; equilibrium point B is a saddle point.

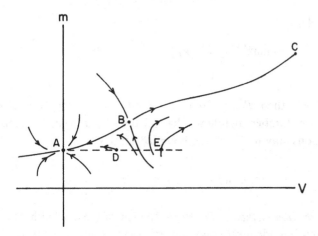

Figure 6.12.

Equilibrium point A is the resting point. Stimulation by a current shock of the correct sign causes a displacement from A to the right on the horizontal dashed line. If the displacement is, say, to the point D, then this is a subthreshold shock and the phase point in the Vm plane, which describes the condition of the axon, returns to the neighborhood of point A. If the displacement reaches the point E, then the phase point proceeds to a neighborhood of equilibrium point C. This describes the first part of an action potential. (The point does not return to equilibrium point A because the variables h and n, which describe how this return takes place, are kept constant.)

The next step is to allow h and n to vary in the way governed by their differential equations

$$\frac{dh}{dt} = \frac{h_\infty(V) - h}{\tau_h(V)},$$

$$\frac{dn}{dt} = \frac{n_\infty(V) - n}{\tau_n(V)},$$

and to study numerically how the equilibrium points of (V, m) change as h and n are varied. The equilibrium points B and C

approach each other, merge, and disappear. The only remaining equilibrium point is A, and the phase point returns to a neighborhood of A. A more detailed account of these changes is given by FitzHugh (1969). Our discussion is not rigorous, and even a more detailed description is numerical and suggestive rather than rigorous. However, this kind of intuitive procedure provides very useful guidance in the study of these complicated nonlinear systems of ordinary differential equations.

The singular perturbation viewpoint. Next we reformulate the HH equations as a singularly perturbed system in a manner parallel to the study of the FitzHugh–Nagumo equations that was described in an earlier section of this chapter. In order to formulate the HH equations as a singularly perturbed system, let

$$T_M(V) = \frac{\tau_m(V)}{\max \tau_m(V)},$$

$$T_H(V) = \frac{\tau_h(V)}{\max \tau_h(V)},$$

$$T_N(V) = \frac{\tau_n(V)}{\max \tau_n(V)}.$$

Then the HH equations become

$$\frac{dV}{dt} = -\frac{1}{C} \left\{ \bar{g}_{Na} m^3 h (V - V_{Na}) + g_K n^4 (V - V_K) + \bar{g}_L (V - V_L) \right\},$$

$$\frac{dm}{dt} = \frac{m_\infty(V) - m}{T_M(V) \max \tau_m(V)},$$

$$\frac{dh}{dt} = \frac{h_\infty(V) - h}{T_H(V) \max \tau_h(V)},$$

$$\frac{dn}{dt} = \frac{n_\infty(V) - n}{T_N(V) \max \tau_n(V)}.$$

Since $\tau_m(V)$ is roughly one-tenth as large as $\tau_h(V)$ and $\tau_n(V)$ [see

FitzHugh (1969), p. 25], we rewrite these equations as

$$\frac{dV}{dt} = -\frac{1}{C}\left\{ \bar{g}_{\mathrm{Na}}m^3h(V-V_{\mathrm{Na}}) + \bar{g}_{\mathrm{K}}n^4(V-V_{\mathrm{K}}) + \bar{g}_L(V-V_L)\right\},$$

$$\frac{dm}{dt} = \frac{m_\infty(V)-m}{T_M(V)},$$

$$\frac{dh}{dt} = \varepsilon\left\{ \frac{h_\infty(V)-h}{T_H(V)}\right\}, \tag{6.13}$$

$$\frac{dn}{dt} = \varepsilon\left\{ \frac{n_\infty(V)-n}{T_N(V)}\right\},$$

where ε is a small positive number (actually about one-tenth). Now we define τ as $\tau = \varepsilon t$. Then $dV/dt = (dV/d\tau)(d\tau/dt) = \varepsilon(dV/d\tau)$ and similarly for dm/dt, dh/dt, and dn/dt. Hence, system (6.13) may be written

$$\varepsilon\frac{dV}{dt} = -\frac{1}{C}\left\{ \bar{g}_{\mathrm{Na}}m^3h(V-V_{\mathrm{Na}}) + \bar{g}_{\mathrm{K}}n^4(V-V_{\mathrm{K}}) + \bar{g}_L(V-V_L)\right\},$$

$$\varepsilon\frac{dm}{dt} = \frac{m_\infty(V)-m}{T_M(V)},$$

$$\varepsilon\frac{dh}{dt} = \varepsilon\left\{ \frac{h_\infty(V)-h}{T_H(V)}\right\},$$

$$\varepsilon\frac{dn}{dt} = \varepsilon\left\{ \frac{n_\infty(V)-n}{T_N(V)}\right\}$$

or

$$\frac{dV}{dt} = -\frac{1}{\varepsilon C}\left\{ \bar{g}_{\mathrm{Na}}m^3h(V-V_{\mathrm{Na}}) + \bar{g}_{\mathrm{K}}n^4(V-V_{\mathrm{K}}) + \bar{g}_L(V-V_L)\right\},$$

$$\frac{dm}{dt} = \frac{1}{\varepsilon}\left\{ \frac{m_\infty(V)-m}{T_M(V)}\right\},$$

$$\frac{dh}{dt} = \frac{h_\infty(V)-h}{T_H(V)}, \tag{6.14}$$

$$\frac{dn}{dt} = \frac{n_\infty(V)-n}{T_N(V)}.$$

System (6.14) is a singularly perturbed system and hence the singular perturbation theory of Chapter 5 can be applied to it. Since there are two singularly perturbed equations in system (6.14), then the theory of Sibuya and the theory of Mishchenko and Rozov are applicable. Such an application has not actually been carried out, but seems to be a step that should be taken.

Now we turn to the question of a mathematical description of repetitive firing of the nerve. We have already seen, in the analysis of the FitzHugh–Nagumo equations, how the addition of a constant term I can cause the appearance of a periodic solution. It may be possible, by extending the singular perturbation analysis of the FitzHugh–Nagumo equations described in Section 6.2.1.2, to show that if I is a constant in a suitable interval of values, then the resulting HH equations have a periodic solution. It should be observed that the two-dimensional case (illustrated in Fig. 6.7) suggests that if I exceeds a fixed value, then a discontinuous periodic solution appears and the amplitude of the discontinuous solution at once exceeds a fixed value. That is, the amplitude does not increase continuously from zero as would be the case with a Hopf bifurcation.

The Hopf bifurcation theorem has been used by Troy (1974/75) to investigate the appearance of periodic solutions as the current I is varied. In order to determine which of the two approaches, the singular perturbation method or the Hopf bifurcation theorem, is more realistic, it is enlightening to study the numerical analysis [carried out by Cole et al. (1955)].

In the graph given by Cole et al. (1955, p. 166), it is shown that if the current I takes larger values, first one action potential occurs, then for a larger value of I, there are three action potentials, and finally, if I is large enough, the action potentials continue indefinitely, that is, an oscillatory phenomenon occurs. In terms of the mathematical description, there is a periodic solution. However, one point should be emphasized about the graphs just described. All the action potentials have the same amplitude. Unlike the result of the Hopf bifurcation theorem, where, as the parameter changes, the amplitude of the periodic solution increases monotonically from zero upward, the periodic solution, as soon as it appears, has a fixed amplitude. Thus, the numerical experimental result is more like the appearance of a periodic solution in a singularly perturbed system, as illustrated for the two-dimensional case in Fig. 6.6.

6.2.2   Analysis of the Noble model of the cardiac Purkinje fiber
6.2.2.1   *The Noble equations viewed as a singularly perturbed system*

As the earlier discussion (Chapter 4) shows, a mathematical analysis of the Noble model should account for or describe two phenomena: the transmission of impulses that originate in the pacemaker region of the heart and the regular spontaneous firing of the Purkinje fibers. The regular spontaneous firing corresponds to a stable periodic solution of the model, and the transmission of impulses would probably be well described by the mathematical theory of entrainment of frequency described in Chapter 5, as we will see later.

However, before undertaking the search for a periodic solution, we must first choose a viewpoint of the Noble equations. In studying the FitzHugh–Nagumo equations qualitatively, it turned out to be very enlightening to regard them as a two-dimensional singularly perturbed system. We shall find that a similar viewpoint is helpful in studying the Noble equations. We first look at a few numerical values to guide our choice of a viewpoint.

The Noble equations have the following form:

$$\frac{dV}{dt} = -\frac{1}{C_m}\{(400m^3h + 0.14)(V - 40) + (1.2\exp[(-V - 90)|50]$$

$$+ 0.015\exp[(V + 90)|60] + 1.2n^4)(V + 100)\},$$

$$\frac{dm}{dt} = \alpha_m(1 - m) - \beta_m m,$$

$$\frac{dh}{dt} = \alpha_h(1 - h) - \beta_h h, \tag{6.15}$$

$$\frac{dn}{dt} = \alpha_n(1 - n) - \beta_n n,$$

where V denotes the membrane potential (inside potential minus outside potential) and we are assuming that the anion current I_{An} is zero. [Noble (1962, pp. 319 and 343ff) discusses the effect of varying I_{An}. We choose $I_{An} = 0$ because this is a value for which Noble obtained numerical results and we can compare our qualitative results with the numerical results.] The other functions that appear in the equations are defined as follows:

$$\alpha_m = \frac{0.1(-V-48)}{\exp[(-V-48)/15]-1},$$

$$\beta_m = \frac{0.12(V+8)}{\exp[(V+8)/5]-1},$$

$$\alpha_h = 0.17\exp\left(\frac{-V-90}{20}\right),$$

$$\beta_h = \left[\exp\left(\frac{-V-42}{10}\right)+1\right]^{-1},$$

$$\alpha_n = \frac{0.0001(-V-50)}{\exp[(-V-50)/10]-1},$$

$$\beta_n = 0.002\exp\left(\frac{-V-90}{80}\right).$$

Following the same convention used in the study of the Hodgkin–Huxley equations, we may write the last three equations of system (6.15) as

$$\frac{dm}{dt} = \frac{m_\infty(V)-m}{\tau_m(V)},$$

$$\frac{dh}{dt} = \frac{h_\infty(V)-h}{\tau_h(V)},$$

$$\frac{dn}{dt} = \frac{n_\infty(V)-n}{\tau_n(V)},$$

where

$$m_\infty(V) = \frac{\alpha_m(V)}{\alpha_m(V) + \beta_m(V)},$$

$$\tau_m(V) = \frac{1}{\alpha_m(V) + \beta_m(V)},$$

and $h_\infty(V)$, $\tau_h(V)$, $n_\infty(V)$, and $\tau_n(V)$ are similarly defined. Numerical values of the functions $\tau_m(V)$, $\tau_h(V)$, and $\tau_n(V)$ are listed in Table 6.1. Notice that

$$\min_V \tau_m(V) \approx 0.1, \qquad \min_V \tau_h(V) \approx 1, \qquad \min_V \tau_n(V) \approx 106,$$

$$\overline{M}_m = \max_V \tau_m(V) \approx 0.206, \qquad \overline{M}_h = \max_V \tau_h(V) \approx 8.3,$$

$$\overline{M}_n = \max_V \tau_n(V) \approx 535,$$

$$\min_V \frac{\tau_m(V)}{\overline{M}_m} \approx 0.486, \qquad \min_V \frac{\tau_h(V)}{\overline{M}_n} \approx 0.121,$$

$$\min_V \frac{\tau_n(V)}{\overline{M}_n} \approx 0.198.$$

These values suggest that $\tau_h(V)$ is between 5 and 10 times as large

Table 6.1

V	$\tau_m(V)$	$\tau_h(V)$	$\tau_n(V)$
-90	0.09889	5.6129	481.9277
-80	0.11025	8	520.833
-70	0.12346	8.3472	534.759
-60	0.13852	5.5586	510.204
-50	0.15528	3.003	452.488
-40	0.17302	1.77399	377.358
-30	0.190476	1.287	306.748
-20	0.204499	1.10485	250.62657
-10	0.20627	1.03745	207.90
0	0.19058	1.013	176.056
10	0.167196	1.0044	151.97568
20	0.14522	1.0014	132.9787
30	0.12746	1.0003	118.483
40	0.112209	1.00005	106.496

as $\tau_m(V)$, and $\tau_n(V)$ is more than 60 times as large as $\tau_h(V)$ and more than 500 times as large as $\tau_m(V)$. If we let

$$M_m = \mathrm{lub}\, \tau_m(V),$$

$$M_h = \mathrm{lub}\, \tau_h(V),$$

$$M_n = \mathrm{lub}\, \tau_n(V),$$

where the lub is taken over V in the interval $[-90, 40]$ (this interval contains the set of values actually assumed by the membrane potential), and if we define

$$T_m(V) = \tau_m(V)/M_m,$$

$$T_h(V) = \tau_h(V)/M_h,$$

$$T_n(V) = \tau_n(V)/M_n,$$

then system (6.15) may be rewritten

$$\frac{dV}{dt} = \mathscr{F}(V, m, h, n),$$

$$\frac{dm}{dt} = \frac{m_\infty(V) - m}{M_m T_m(V)},$$

$$\frac{dh}{dt} = \frac{h_\infty(V) - h}{M_h T_h(V)}, \qquad\qquad (6.16)$$

$$\frac{dn}{dt} = \frac{n_\infty(V) - n}{M_n T_n(V)},$$

where, for convenience, we have written the right-hand side of the first equation in system (6.15) in abbreviated form. The functions $T_m(V)$, $T_h(V)$, and $T_n(V)$ are all of the same order of magnitude in that their values are positive numbers in the interval $[0.1, 1]$. The numerical values previously displayed of the functions $\tau_m(V)$ and $\tau_h(V)$ suggest that M_m is about 0.21, M_h is about 8.3, and M_n is about 535. Thus, M_h is about $40M_m$ and M_n is about $2548M_m$.

Hence, we can rewrite (6.16) as

$$\frac{dV}{dt} = \mathscr{F}(V, m, h, n),$$

$$\frac{dm}{dt} = \frac{m_\infty(V) - m}{M_m T_m(V)},$$

$$\frac{dh}{dt} = \frac{h_\infty(V) - h}{40 M_m T_h(V)}, \tag{6.17}$$

$$\frac{dn}{dt} = \frac{n_\infty(V) - n}{2548 M_m T_n(V)}.$$

Further, if we let $\varepsilon = 1/40$ and $\eta = 1/(2548/40) \approx 1/64$, then we may rewrite (6.17) as

$$\frac{dV}{dt} = \mathscr{F}(V, m, h, n),$$

$$\frac{dm}{dt} = \frac{m_\infty(V) - m}{M_m T_m(V)},$$

$$\frac{dh}{dt} = \varepsilon \left[\frac{h_\infty(V) - h}{M_m T_h(V)} \right], \tag{6.18}$$

$$\frac{dn}{dt} = \varepsilon \eta \left[\frac{n_\infty(V) - n}{M_m T_n(V)} \right].$$

Now we introduce a conventional change of the independent variable; that is, we set

$$\tau = \varepsilon t,$$

where ε is regarded as a small positive number. (Since ε is small, this change of variable amounts to "stretching out" the time axis.) Then by the chain rule from calculus,

$$\frac{dV}{dt} = \frac{dV}{d\tau} \frac{d\tau}{dt} = \frac{dV}{d\tau} \varepsilon,$$

$$\frac{dm}{dt} = \frac{dm}{d\tau} \varepsilon,$$

$$\frac{dh}{dt} = \frac{dh}{d\tau} \varepsilon,$$

$$\frac{dn}{dt} = \frac{dn}{d\tau} \varepsilon,$$

and we may rewrite (6.18) as

$$\frac{dV}{d\tau}\varepsilon = \mathscr{F}(V, m, h, n),$$

$$\frac{dm}{d\tau}\varepsilon = \frac{m_\infty(V) - m}{M_m T_m(V)},$$

$$\frac{dh}{d\tau}\varepsilon = \varepsilon\left[\frac{h_\infty(V) - h}{M_m T_h(V)}\right],$$

$$\frac{dn}{d\tau}\varepsilon = \varepsilon\eta\left[\frac{n_\infty(V) - n}{M_m T_n(h)}\right]$$

or, denoting the independent variable by t so as to conform to conventional notation, we have

$$\frac{dV}{dt} = \frac{1}{\varepsilon}\mathscr{F}(V, m, h, n),$$

$$\frac{dm}{dt} = \frac{1}{\varepsilon}\left[\frac{m_\infty(V) - m}{M_m T_m(V)}\right],$$

$$\frac{dh}{dt} = \frac{h_\infty(V) - h}{M_m T_h(V)}, \tag{6.19}$$

$$\frac{dn}{dt} = \eta\left[\frac{n_\infty(V) - n}{M_m T_n(V)}\right].$$

Now ε and η were originally equal to $1/40$ and $1/64$, respectively. However, if we regard ε and η as small positive parameters, then the original problem of mathematically analyzing the solutions of system (6.15) suggests the following problem: to analyze the solutions of (6.19) when ε and η are sufficiently small positive numbers.

Notice that we have not said that this new problem is strictly equivalent to the original problem of studying system (6.15). To begin with, in deriving (6.19) we used the approximations

$$M_h = 40 M_m,$$
$$M_n = 2548 M_m.$$

Second, even if we are able to make a rigorous analysis of (6.19) for sufficiently small positive ε and η, it does not follow that the conclusions are applicable to (6.15) unless we can show that the rigorous analysis of (6.19) is applicable to a range of values of ε and η that includes $\varepsilon = 1/40$ and $\eta = 1/64$. As we have already seen in Chapter 5, theorems that yield results valid for sufficiently small ε often do not give explicit estimates on the range of validity of ε.

In spite of these qualifications or reservations, we shall go ahead to the analysis of the solutions of (6.19). There are two reasons for this decision. First, as seen earlier in this chapter, the analysis of a similar approximation of the FitzHugh–Nagumo equations turned out to yield a lot of enlightening and useful information about behavior of solutions. Second, there is the rather negative reason that a direct rigorous analysis of the solutions of (6.15) seems to be beyond the reach of known methods, whereas we can obtain at least some partial results concerning solutions of (6.19) if ε and η are sufficiently small positive numbers.

6.2.2.2 *Analysis of the singularly perturbed system*

Reduction to three-dimensional system. In view of the messiness of the function (V, m, h, n) in the first equation of system (6.19), one might expect that the chief difficulties that would arise in the study of (6.19) would stem from the fact that one is dealing with complicated algebraic expressions. Actually this is not so. It turns out that the beginnings of a qualitative analysis of (6.19) are readily accessible. The obstacle to a complete qualitative analysis lies in the absence of enough qualitative theory about singularly perturbed systems.

Now we turn to the analysis of (6.19). First, let us compare system (6.19) with system (6.14), the version of the Hodgkin–Huxley equations that is obtained in the preceding section and that is a singularly perturbed system. Systems (6.14) and (6.19) are similar in that both are four-dimensional singularly perturbed systems and each has a two-dimensional "fast" system [the first two equations in each of the systems (6.14) and (6.19)]. There is, however, an important difference between (6.14) and (6.19). In (6.19), the last equation has a small parameter η multiplying the right-hand side. Hence, the last equation can be regarded as a perturbation of the equation

$$\frac{dn}{dt} = 0.$$

The conventional mathematical approach to such a system is to assume that $\eta = 0$. Then $dn/dt = 0$ and $n(t)$ must be a constant, say n_0, and system (6.19) becomes

$$\frac{dV}{dt} = \frac{1}{\varepsilon}\mathscr{F}(V, m, h, n_0),$$

$$\frac{dm}{dt} = \frac{1}{\varepsilon}\frac{m_\infty(V) - m}{M_m T_m(V)},$$

$$\frac{dh}{dt} = \frac{h_\infty(V) - h}{M_m T_h(V)}.$$

(6.20)

The solutions of (6.20) are then studied, and it is shown that the solutions of (6.19) with η nonzero but small are close to the solutions of (6.20).

[Note that although (6.20) is a singularly perturbed system, it is a three-dimensional system. Consequently if the conventional mathematical approach just sketched can be applied, we would expect the analysis of (6.19) to be easier than the analysis of (6.14), the singular perturbation version of the Hodgkin–Huxley equations, because (6.14) is a four-dimensional system.]

Our first step will consist in showing that this conventional mathematical approach fails. The approach fails for two reasons. The first reason for failure is purely mathematical: The analysis of (6.20) shows that it has no periodic solutions. Thus, we have no hope of using this approach to find a periodic solution of (6.19). Although this is a negative result, we will carry out a detailed analysis of (6.20) in the next section because the results will be referred to in a later section and because it provides an application of the singular perturbation theory introduced in Chapter 5.

The second reason for failure is more fundamental: The assumption $dn/dt = 0$ is physiologically untenable. This point will be discussed in more detail following the analysis of (6.20).

Demonstration that the three-dimensional system does not have periodic solutions. Since we are looking for a periodic solution that would describe the regular spontaneous firing of the Purkinje fiber, a reasonable procedure is to use the singular perturbation theory to establish a periodic solution, with orbit, say $\mathscr{S}$, of (6.20), and then show that if η is sufficiently small, system (6.19) has a periodic solution or an approximately periodic solution whose orbit is close to the orbit $\mathscr{S}$. However, we shall now show that this seemingly reasonable procedure fails, that is, we will show that we do not obtain a periodic solution for system (6.20). Then we will explain why the physiology would lead us to expect this result.

Following the theory of singularly perturbed systems described in Chapter 5, we seek discontinuous solutions of the degenerate system

$$\mathscr{F}(V, m, h, n_0) = 0,$$

$$\frac{m_\infty(V) - m}{M_m T_m(V)} = 0,$$

$$\frac{dh}{dt} = \frac{h_\infty(V) - h}{M_m T_h(V)}$$

or

$$\mathscr{F}(V, m, h, n_0) = 0,$$

$$m = m_\infty(V),$$

$$\frac{dh}{dt} = \frac{h_\infty(V) - h}{M_m T_h(V)}.$$

A discontinuous solution is a curve consisting of two kinds of segments:

(i) those that are described by solutions of

$$\frac{dh}{dt} = \frac{m_\infty(V) - h}{M_m T_h(V)}$$

and that are contained in the manifold $\mathscr{M}$ defined by

$$\mathscr{M} = \{(V, m, h)/\mathscr{F}(V, m, h, n_0) = 0 \text{ and } m = m_\infty(V)\};$$

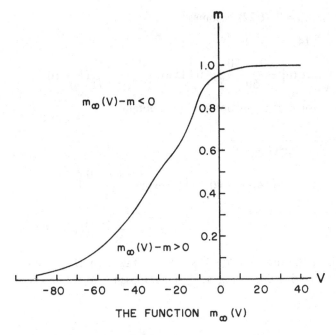

Figure 6.13. The function $m_\infty(V)$.

(ii) those segments that are "jump arcs," that is, solutions of the fast system.

To visualize the discontinuous solution clearly, we obtain first a picture of the manifold $\mathcal{M}$. The manifold $\mathcal{M}$ is described by the two equations

$$m = m_\infty(V), \tag{6.21}$$

$$\mathcal{F}\big(V, m_\infty(V), h, n_0\big) = 0. \tag{6.22}$$

Equation (6.21) describes a cylindrical surface that is perpendicular to the (V, m) plane. This surface is simple to visualize because the function $m_\infty(V)$ is the simple monotonic increasing function graphed in Fig. 6.13 [see Noble (1962, p. 326, Fig. 4)]. Equation (6.22) describes a cylindrical surface perpendicular to the (V, h) plane. Although the function $\mathcal{F}$ is complicated looking, the curve in the (V, h) plane described by (6.22) is a simple configuration.

For the case $n_0 = 0$, (6.22) becomes

$$(400m^3h + 0.14)(V - 40)$$

$$= -\left\{1.2\exp\left(\frac{-V - 90}{50}\right) + 0.015\exp\left(\frac{V + 90}{60}\right)\right\}(V + 100),$$

and solving for h as a function of V, we obtain

$$h = h(V) = \frac{1}{400m^3(V - 40)}$$

$$\times\left\{-(0.14)(V - 40) - \left[1.2\exp\left(\frac{-V - 90}{50}\right)\right.\right.$$

$$\left.\left. + 0.015\exp\left(\frac{V + 90}{60}\right)\right](V + 100)\right\}.$$

The graph of the function $h(V)$ is given in Fig. 6.14. [Because of the scale of the drawing in Fig. 6.14, it is impossible to see how the

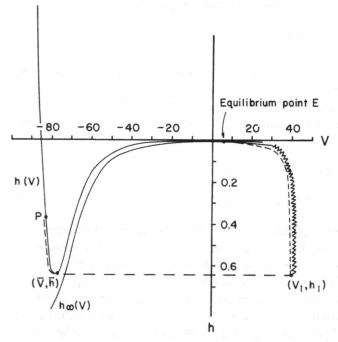

Figure 6.14.

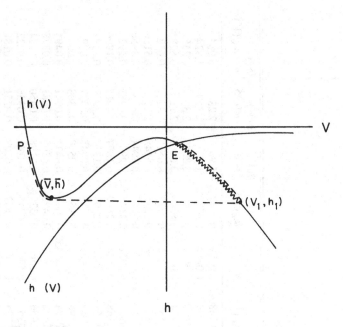

Figure 6.15.

graphs of the functions $h(V)$ and $h_\infty(V)$ intersect. A distorted version of these graphs, which illustrates the intersection more clearly, is shown in Fig. 6.15.] It is essential to observe that the graph of $h_\infty(V)$ intersects the graph of $h(V)$ at the point E, where $h(V)$ is an *increasing* function. The graph of the function $h(V)$ drawn in Fig. 6.14 is obtained by plotting the points whose coordinates are listed in Table 6.2 and joining these points by a smooth curve. The in-the-large properties of the resulting curve, especially the property of being S-shaped, play an important role in determining the form of the discontinuous solutions, as we will see shortly. One might well argue that if more points on the graph of the function were plotted, then the in-the-large character of the graph and, hence, the form of the discontinuous solutions could be significantly changed. The answer to this is that the functions that appear in the original Noble equations are chosen to be consistent with a finite set of data points obtained from laboratory observations. That is, the equations are empirical equations or are approximate summaries of laboratory data. Hence, a small change in

Table 6.2

V	α_m	β_m	α_n	β_n	α_h	β_h	$m_\infty(V)$	$n_\infty(V)$	$h_\infty(V)$	$h(V)$ $n=0$	$h(V)$ $n=0.75$	$h(V)$ $n=n_\infty(V)$
−90	0.272	9.84	0.000075	0.002	0.17	0.00816	0.02689	0.0359	0.9542	−5.98	−2.23	−5.984
−87	0.3129	9.48	0.000094	0.00193	0.1463	0.0109	0.0309	0.0464	0.9306	−1.92	1.369	−1.924
−83	0.3758	9.00	0.00013	0.00183	0.1197	0.0163	0.0400	0.0644	0.8800	0.2545	2.3	0.255
−80	0.4299	8.64	0.00016	0.00176	0.1031	0.0219	0.0474	0.0817	0.8248	0.6267	2.11	0.627
−77	0.4905	8.28	0.00019	0.0017	0.0887	0.0293	0.0559	0.1024	0.7517	0.6519	1.72	0.652
−73	0.5821	7.80	0.00026	0.00162	0.0726	0.0431	0.0694	0.1368	0.6275	0.5148	1.193	0.5156
−70	0.6597	7.44	0.00031	0.00156	0.0625	0.0573	0.0814	0.1673	0.5217	0.3944	0.874	0.3956
−67	0.7454	7.08	0.00038	0.0015	0.0538	0.0759	0.0953	0.2021	0.4148	0.29	0.6283	0.292
−63	0.8729	6.60	0.00049	0.00142	0.0441	0.109	0.1168	0.2544	0.2880	0.1877	0.4017	0.1906
−60	0.9790	6.24	0.00058	0.00138	0.0379	0.142	0.1356	0.2978	0.2107	0.1337	0.2859	0.1375
−57	1.095	5.88	0.00069	0.00132	0.0326	0.182	0.1569	0.3426	0.1519	0.0948	0.2037	0.0995
−53	1.264	5.40	0.00085	0.00126	0.0267	0.2497	0.1896	0.405	0.0966	0.0599	0.1303	0.066
−50	1.40	5.04	0.001	0.00121	0.0230	0.31	0.2174	lim = 0.452	0.0691	0.0427	0.0941	0.0495
−47	1.551	4.68	0.00116	0.00117	0.0198	0.3775	0.2488	0.4983	0.0498	0.0305	0.0681	0.0378
−43	1.765	4.20	0.00139	0.00111	0.0162	0.475	0.2956	0.5568	0.0329	0.0198	0.0450	0.0274
−40	1.93	3.85	0.00158	0.00107	0.0139	0.5498	0.3351	0.5960	0.0246	0.0145	0.0336	0.0222
−37	2.117	3.49	0.00179	0.00103	0.0120	0.622	0.3775	0.6345	0.0189	0.0107	0.0251	0.0181
−33	2.373	3.02	0.00208	0.00098	0.0098	0.7109	0.4400	0.6796	0.0136	0.0073	0.0175	0.0142
−30	2.58	2.67	0.00231	0.00095	0.0085	0.7685	0.4911	0.7097	0.0109	0.0055	0.0135	0.0119

−27	2.787	2.33	0.00056	0.00091	0.0073	0.8175	0.5444	0.7378	0.0088	0.0043	0.0082	0.0103
−23	3.082	1.89	0.00289	0.00087	0.0059	0.8698	0.6194	0.7695	0.0067	0.0032	0.0080	0.0085
−20	3.31	1.58	0.00316	0.00083	0.0051	0.9000	0.6763	0.7912	0.0056	0.0026	0.0067	0.0076
−17	3.55	1.294	0.00343	0.00080	0.0044	0.9241	0.7329	0.8103	0.0047	0.0022	0.0057	0.0069
−13	3.876	0.949	0.00379	0.00076	0.0036	0.9478	0.8033	0.8322	0.0038	0.0018	0.0048	0.0063
−10	4.12	0.728	0.00407	0.00074	0.0031	0.9608	0.8498	0.8469	0.0032	0.0016	0.0044	0.0062
−7	4.385	0.542	0.00436	0.00071	0.0027	0.9707	0.8899	0.8602	0.00277	0.0015	0.0042	0.0061
−3	4.736	0.379	0.00474	0.00067	0.0022	0.9801	0.9314	0.8755	0.00244	0.00148	0.0041	0.0064
0	5.004	0.243	0.00503	0.00065	0.0019	0.9852	0.9539	0.8857	0.00192	0.00151	0.0042	0.0068
3	5.276	0.164	0.00533	0.00063	0.0016	0.989	0.9699	0.895	0.00162	0.00158	0.0045	0.0075
7	5.644	0.094	0.00572	0.00059	0.0013	0.9926	0.9836	0.9057	0.00131	0.0017	0.0049	0.0086
10	5.92	0.061	0.00601	0.00057	0.0011	0.9945	0.9898	0.9130	0.0011	0.0019	0.0055	0.0048
13	6.206	0.038	0.00631	0.00055	0.0009	0.9959	0.9939	0.9156	0.0009	0.0022	0.0062	0.0113
17	6.586	0.020	0.00671	0.00053	0.0008	0.9973	0.9969	0.9274	0.0008	0.0026	0.0075	0.0139
20	6.874	0.012	0.00701	0.00051	0.0007	0.9979	0.9983	0.9327	0.0007	0.0031	0.0088	0.0168
23	7.163	0.0075	0.00731	0.00049	0.0006	0.9985	0.9989	0.9375	0.0006	0.0037	0.0106	0.0205
27	7.551	0.0038	0.00771	0.00046	0.0005	0.9989	0.9995	0.9433	0.0005	0.0051	0.0143	0.0283
30	7.843	0.0023	0.008	0.00044	0.0004	0.9993	0.9997	0.9472	0.0004	0.0068	0.0191	0.0382
33	8.137	0.0014	0.0083	0.00043	0.00036	0.9994	0.9998	0.9508	0.00036	0.01	0.0281	0.0567
37	8.529	0.00066	0.0087	0.00041	0.0003	0.9996	0.9999	0.9551	0.0003	0.025	0.0680	0.1387
40	8.825	0.00039	0.009	0.00039	0.00025	0.9997	0.9999	0.9581	0.00025	—	—	—
43	9.121	0.00023	0.0093	0.00038	0.0002	0.9998	0.9999	0.9608	0.0002	−0.0267	−0.072	−0.1486
47	9.517	0.00011	0.0097	0.00036	0.00015	0.9999	0.9999	0.9641	0.00015	−0.0122	−0.032	−0.0665
50	9.814	0.00006	0.01	0.00035	0.0001	0.9999	0.9999	0.9664	0.0001	−0.0088	−0.023	−0.0482

the equations may make them no less accurate, and may indeed make them a more accurate description of the data. Consequently, if we make a small change in the equations in order to be able to apply known theory and draw significant inferences about the solution of the equations, the equations thus modified may provide an equally accurate and possibly more effective model. We shall use this viewpoint a number of times in the analysis that follows.

The graphs in Figs. 6.13 and 6.14 show that the manifold $\mathcal{M}$ is easily visualized. Now all the segments of a discontinuous solution except the jump arcs, lie in $\mathcal{M}$. To get a complete picture of a discontinuous solution, it is only necessary to determine the possible jump arcs. Inspection of the graph in Fig. 6.14 shows that the point $(\bar{V}, m_\infty(\bar{V}), \bar{h}) = (\bar{V}, \bar{m}, \bar{h})$, where $\bar{h} = h(\bar{V})$ and the function $h(V)$ has a maximum at $V = \bar{V}$, is a possible junction point. Then if there is a jump arc starting at this junction point, it is a curve in the plane $h = \bar{h}$. This curve must be an orbit of a solution of the fast system

$$\frac{dV}{dt} = \frac{1}{\varepsilon}\mathscr{F}(V, m, \bar{h}, n_0),$$

$$\frac{dm}{dt} = \frac{1}{\varepsilon}\left[\frac{(m_\infty(V) - m)}{M_m T_m(V)}\right].$$

Since we are looking only at orbits and since the factor $1/\varepsilon$ appears in both equations, then it suffices to look at the orbits of the system

$$\frac{dV}{dt} = \mathscr{F}(V, m, \bar{h}, n_0),$$

$$\frac{dm}{dt} = \frac{m_\infty(V) - m}{M_m T_m(V)}. \tag{6.23}$$

To show that $(\bar{V}, \bar{m}, \bar{h})$ is a junction point, it is necessary first to show that there is exactly one orbit [of a solution of (6.23)] that "comes out" of $(\bar{V}, \bar{m})$. More precisely, it is necessary to prove that there exists a solution $(V(t), m(t))$ of (6.23) such that

$$\lim_{t \to -\infty} (V(t), M(t)) = (\bar{V}, \bar{m})$$

and that all solutions with this property have the same orbit. Let us assume that such a solution with orbit, say, $\mathcal{S}_1$, exists. (It is straightforward but fairly lengthy to prove this assumption.)

As can be visualized from Fig. 6.14, system (6.23) has just two equilibrium points: $(\overline{V}, \overline{m})$, where $\overline{m} = m_\infty(\overline{V})$, and another point that we denote by (V_1, m_1), where V_1 is the positive solution of the equations $h = h(V)$ and $h = h(\overline{V})$, and $m_1 = m_\infty(V_1)$. [The point (V_1, h_1), where $h_1 = h(V_1)$, is shown in Fig. 6.14.] In order to show that $\mathcal{S}_1$ is a jump arc, we must show that orbit $\mathcal{S}_1$ has as its ω-limit set the point (V_1, m_1). In order to show this, we observe that

$$\mathscr{F}\left(V, m, \overline{h}, n_0\right) = -\frac{1}{C_m}\left[400m^3(V - 40)\right]\left[\overline{h} - h(V)\right].$$

In this discussion, we assume $V < 40$. Hence, $V - 40 < 0$. From the description of the function $m_\infty(V)$ given by Noble (1962, see especially p. 326, Fig. 4) and the equation

$$\frac{dm}{dt} = \frac{m_\infty(V) - m}{M_m T_m(V)},$$

it follows that if the initial value of $m(t)$ is in the interval $(0, 1)$ (and this is the only case that is physiologically significant), then for all later t,

$$0 < m(t) < 1.$$

Also since $C_m > 0$, it follows that $\mathscr{F}(V, m, \overline{h}, n_0)$ has the same sign as $\overline{h} - h(V)$, but $\overline{h}$ is a local maximum value of $h(V)$. This can be seen from the graph in Fig. 6.14. (The positive direction on the h axis in that graph is downward.) Hence, it follows that

$$\overline{h} - h(V) > 0$$

for $V < \overline{V}$ and for $V \in (\overline{V}, V_1)$. It can also be seen that

$$\mathscr{F}\left(\overline{V}, m, \overline{h}, n_0\right) = \mathscr{F}\left(V_1, m, \overline{h}n_0\right) = 0$$

and that, if $V > V_1$,

$$\left(V, m, \overline{h}, n_0\right) < 0.$$

We have already observed that

$$0 < m(t) < 1$$

and inspection of $\mathscr{F}(V, m, \bar{h}, n_0)$ shows that if V is a sufficiently large positive number, dV/dt is negative, and if V is a negative number of sufficiently large absolute value, then dV/dt is positive. Hence, $V(t)$ remains bounded for all t. Hence, the solution with orbit $\mathscr{S}_1$ remains bounded. Moreover, it follows from the signs of $\mathscr{F}(V, m, \bar{h}, n_0)$ and $[m_\infty(V) - m]$ that the solution with orbit $\mathscr{S}_1$ must have as its ω-limit set an equilibrium point, that is, either (V, m) or $(\bar{V}_1, \bar{m}_1)$. But a solution of system (6.23) that "comes out" of $(\bar{V}, \bar{m})$ must be in the set

$$A = \{(V, m)/\bar{V} < V < V_1\}.$$

Since $\mathscr{F}(V, m, \bar{h}, n_0) > 0$ on the set A, then dV/dt is positive. Therefore, the orbit $\mathscr{S}_1$ must approach the equilibrium point (V_1, m_1). Hence, the orbit $\mathscr{S}_1$ may be regarded as a jump arc. Now the next segment of the discontinuous solution would start at the point $(V_1, m_1, \bar{h})$, where $\bar{h} = h(V_1)$ and $m_1 = m_\infty(V_1)$. Thus, this next segment would be a solution of the equation

$$\frac{dh}{dt} = \frac{h_\infty(V) - h}{M_m T_h(V)}$$

with initial value $\bar{h}$. Also, the solution lies in the manifold described by the two equations

$$m = m_\infty(V), \tag{6.24}$$

$$\mathscr{F}(V, m_\infty(V), h, n_0) = 0. \tag{6.25}$$

The graph of $h = h(V)$ in Fig. 6.14 shows that near (V_1, h_1), we can solve (6.25) uniquely for V as a function of h, that is, we obtain

$$V = G(h),$$

and the segment is described by the solution $h(t)$ of the equation

$$\frac{dh}{dt} = \frac{h_\infty[G(h)] - h}{M_h T_h[G(h)]}$$

such that $h(0) = \bar{h}$, by $V = G[h(t)]$ and by

$$m = m_\infty(V) = m_\infty\{G[h(t)]\}.$$

The projection of the segment just described on the (V, h) plane is shown in Fig. 6.14 by a jagged line. However, this solution approaches the equilibrium point E that has coordinate $\bar{\bar{V}}$ (between 0 and ± 10) that is such that $h(\bar{\bar{V}}) = h_\infty(\bar{\bar{V}})$. [Thus E has coordinates $(\bar{\bar{V}}, m_\infty(\bar{\bar{V}}), h_\infty(\bar{\bar{V}}))$.]

The same is true for other discontinuous solutions; that is, each discontinuous solution approaches the point E as $t \to \infty$. For example, consider the discontinuous solution that has as its initial point the point P in Fig. 6.15. [Actually P is the projection of that initial point from manifold $\mathcal{M}$ onto the (V, h) plane.] The projection of this discontinuous solution on the (V, h) plane is indicated by the dashed line.

It follows from the theory of Mishchenko and Rozov that if ε is small, system (6.20) with $n_0 = 0$ has no periodic solutions; that is, in this case, $n_0 = 0$, we obtain no periodic solutions by using the viewpoint of singular perturbation theory. Further analysis of the same kind for other fixed values of n_0 shows that, again, no periodic solution is obtained; that is, for fixed values of n_0, the discontinuous solutions approach an equilibrium point. Thus, the strategy proposed earlier, that is, to find a periodic solution of (6.20) and then show that if η is sufficiently small, system (6.19) has an approximately periodic solution, is doomed to failure. System (6.20) does not have closed discontinuous solutions.

Physiological reasons why three-dimensional system does not have periodic solutions. When we consider the physiological significance of the variable n, it becomes clear that this failure is inevitable. The variable n is the activation variable for the flow of potassium ions. The varying flow of potassium ions produces the pacemaker effect that in turn makes possible the regular firing of the Purkinje fiber [see Noble (1962, p. 333)]. If the activation variable n remains constant, the underlying ideas in the derivation of the Noble equations are no longer valid. [See the discussion in Noble (1962, pp. 326–327). In that discussion, the variation in n plays a crucial role in the repetitive action in the Purkinje fiber.] Thus, there is no possibility of reducing our problem to the study of the three-dimensional system (6.20).

Comparison between four-dimensional and three-dimensional systems. As we have seen, approximating the solutions of (6.19) with solutions of (6.20) fails because (6.20) has no periodic solutions. That is, the periodic solutions of (6.19), if there are any, elude this approach. This is actually not surprising because in searching for periodic solutions, we are looking at the asymptotic behavior of solutions, that is, the behavior of the solutions as $t \to \infty$. The technique of approximating solutions of (6.19) with solutions of (6.20) is valid only for intervals on the t axis of finite length. This is illustrated by the following examples.

Example 1.

$$\frac{dx}{dt} = 0,$$
$$\frac{dy}{dt} = \eta y, \tag{6.26}$$

where η is a small positive number. If $\eta = 0$, then each solution of (6.26) is a constant 2-vector. But if $\eta > 0$, then

$$y(t) = Ke^{\eta t},$$

where K is a constant. Thus, if $K \neq 0$, then

$$\lim_{t \to \infty} |y(t)| = \infty.$$

This is certainly not the behavior of solutions of (6.26) with $y = 0$.

Example 2.

$$\frac{dx}{dt} = 0,$$
$$\frac{dy}{dt} = \eta z, \tag{6.27}$$
$$\frac{dz}{dt} = \eta y,$$

where η is a small positive number. If $\eta = 0$, then each solution of (6.27) is a constant 3-vector. But if $\eta > 0$, then a fundamental matrix of (6.27) is

$$\begin{bmatrix} 1 & 0 & 0 \\ 0 & \cos \eta t & -\sin \eta t \\ 0 & \sin \eta t & \cos \eta t \end{bmatrix}.$$

That is, each solution of (6.27) is a linear combination of the columns of the preceding matrix. The solution

$$\begin{bmatrix} 0 \\ \cos \eta t \\ \sin \eta t \end{bmatrix}$$

is certainly not well approximated by a constant 3-vector no matter how small η is.

Example 2 shows how a periodic solution can elude us if the solutions of the equation with $\eta = 0$ are studied. It is also enlightening to take the singular perturbation viewpoint illustrated by the following example.

Example 3.

$$\frac{dx}{dt} = -\frac{1}{\eta}(x^3 - 3x - y) \quad \text{or} \quad \eta \frac{dx}{dt} = -(x^3 - 3x - y),$$

$$\frac{dy}{dt} = x. \tag{6.28}$$

The degenerate system corresponding to (6.28) is

$$x^3 - 3x - y = 0,$$

$$\frac{dy}{dt} = x.$$

The degenerate system is the result of setting $\eta = 0$, and solving the degenerate system means solving

$$\frac{dy}{dt} = x$$

on the manifold defined by

$$y = x^3 - 3x.$$

Thus, no periodic solution is obtained. However, the same kind of analysis that was used earlier in this chapter shows that (6.28) has a closed discontinuous solution and, hence, a periodic solution.

With these examples in mind, we return to system (6.19). Let us first make a change of the independent variable in (6.19):

$$\tau = \eta t.$$

Then (6.19) becomes

$$\frac{dV}{d\tau} = \frac{dV}{dt}\frac{dt}{d\tau} = \frac{1}{\eta}\frac{1}{\varepsilon}\mathscr{F}(V, m, h, n),$$

$$\frac{dm}{d\tau} = \frac{1}{\eta}\frac{1}{\varepsilon}\frac{m_\infty(V) - m}{M_m T_m(V)},$$

$$\frac{dh}{d\tau} = \frac{1}{\eta}\frac{h_\infty(V) - h}{M_m T_h(V)},$$

$$\frac{dh}{d\tau} = \frac{n_\infty(V) - n}{M_m T_n(V)}.$$

(6.29)

If we regard ε simply as a fixed constant and η as a small parameter, then system (6.29) may be regarded as a singularly perturbed system in which the fast system consists of the first three equations and the slow system is the last equation. Consequently, one segment of a discontinuous solution of (6.29) consists of an orbit of the fast system

$$\frac{dV}{d\tau} = \frac{1}{\eta}\frac{1}{\varepsilon}\mathscr{F}(V, m, h, \bar{n}),$$

$$\frac{dm}{d\tau} = \frac{1}{\eta}\frac{1}{\varepsilon}\frac{m_\infty(V) - m}{M_m T_m(V)},$$

$$\frac{dh}{d\tau} = \frac{1}{\eta}\frac{h_\infty(V) - h}{M_m T_h(V)},$$

(6.30)

where $\bar{n}$ is a constant. But the orbits of (6.30) are the orbits of system (6.20) that was investigated earlier by using singular perturbation analysis (where ε is regarded as a small positive parameter).

Thus, we conclude that a rigorous study of (6.19) might be undertaken by using a two-stage singular perturbation analysis: First, a singular perturbation analysis of (6.30) in which ε is regarded as a small positive parameter; and second, a singular perturbation analysis of (6.29) in which η is a small positive parameter. The results of the analysis of (6.30) are used to carry out the analysis of (6.29).

We have already see how the orbits of (6.20) [and, hence, also of (6.30) with $n = 0$] approach the equilibrium point E. Such an orbit can be regarded as a fast-motion part of a discontinuous solution of (6.29). The fast-motion part is followed by a slow-motion part that is a solution of the degenerate system

$$\frac{dn}{d\tau} = \frac{n_\infty(V) - n}{M_m T_n(V)},$$

$$\mathscr{F}(V, m, h, n) = 0, \qquad \mathscr{F}[V, m_\infty(V), h_\infty(V), n] = 0,$$

$$m = m_\infty(V), \quad \text{or} \quad m = m_\infty(V), \qquad (6.31)$$

$$h = h_\infty(V), \qquad h = h_\infty(V).$$

This solution starts from the point

$$\bar{\bar{P}} = \left(\bar{\bar{V}}, m_\infty(\bar{\bar{V}}), h_\infty(\bar{\bar{V}}), 0\right).$$

[Remember that the point E has coordinates

$$\left(\bar{\bar{V}}, m_\infty(\bar{\bar{V}}), h_\infty(\bar{\bar{V}})\right).]$$

Since $n_\infty(\bar{\bar{V}}) > 0$, then it follows that $dn/d\tau > 0$ at the point $\bar{\bar{P}}$ and, therefore, the solution $n(\tau)$ increases. As $n(\tau)$ increases, V changes in accordance with the equation

$$\mathscr{F}[V, m_\infty(V), h_\infty(V), n(\tau)] = 0.$$

Note that $n(\tau)$ does not "reach" $n_\infty(V)$; that is, there is no number τ_0 such that

$$n(\tau_0) = n_\infty(V_0),$$

where

$$[V_0, m_\infty(V_0, h_\infty(V_0), n(\tau_0)] = 0.$$

The reason for this is that the point

$$(V_0, m_\infty(V_0), h_\infty(V_0), n(\tau_0))$$

would be an equilibrium point of (6.31) and, hence, no orbit could contain that point. Thus, we can only conclude

$$n_\infty(V) - n(\tau)$$

remains positive but is decreasing. To complete the analysis, it would be necessary to determine if a junction point occurs on this slow-motion part, then determine if there exists a drop point

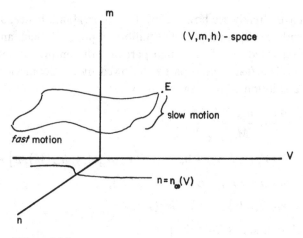

Figure 6.16.

following this junction point, and so forth. This would require, first of all, a fairly extensive numerical study, and we will not undertake such a study here. We observe only that on each slow-motion part of the discontinuous solution, the term

$$n_\infty(V) - n(\tau)$$

is positive and is decreasing as τ increases. Thus, the closed discontinuous solution that would be sought would have the form indicated in Fig. 6.16. [In Fig. 6.16, the (V, m, h) space is represented by the (V, m) plane.]

A modification of the three-dimensional system. We have already indicated the complications that arise in the study of system (6.19). In this section, we describe a three-dimensional system that is a kind of simplification of (6.19), that is, the following modification of system (6.20):

$$\frac{dV}{dt} = \frac{1}{\varepsilon} \mathscr{F}\left[V, m, h, n_\infty(V)\right],$$

$$\frac{dm}{dt} = \frac{1}{\varepsilon}\left[\frac{m_\infty(V) - m}{M_m T_m(V)}\right], \tag{6.32}$$

$$\frac{dh}{dt} = \frac{h_\infty(V) - k}{M_m T_h(V)}.$$

[Formally, this modification of (6.20) consists in replacing n_0 by $n_\infty(V)$ in the first equation.]

There is no rigorous justification for making this modification. Indeed, the geometric behavior of the solutions of system (6.32) is different from the behavior of the kind of periodic solutions that we would expect to get for system (6.19). The fast-motion part of the discontinuous solution approaches E as $t \to \infty$. Thus, for t sufficiently large, V is approximately equal to $\overline{\overline{V}}$. Thus, the fourth equation in system (6.29) [which is system (6.19) with a simple change of the independent variable]) becomes

$$\frac{dn}{dt} = \frac{n_\infty(\overline{\overline{V}}) - n}{M_m T_n(\overline{\overline{V}})},$$

and the solution $n(\tau)$ of this equation with $n(0) = 0$ is easily seen to be

$$n(\tau) = n_\infty(\overline{\overline{V}})\left[1 - \exp\left(\frac{-\tau}{M_m T_n(\overline{\overline{V}})}\right)\right].$$

Thus, $n(\tau) \to n_\infty(\overline{\overline{V}})$ as $\tau \to \infty$. [In actuality, as $n(\tau)$ increases, the value of V, which is governed by the equation

$$\mathscr{F}[V, m_\infty(V), h_\infty(V), n] = 0,$$

also changes.]

By considering the system (6.32), we are assuming in advance that $n = n_\infty(V)$. The graph of a solution of (6.32) would be represented in Fig. 6.16 by a curve in the surface

$$n = n_\infty(V)$$

in (V, m, h, n) space.

By applying the Mishchenko–Rozov theory, a detailed analysis of system (6.32) can be made, and existence of a periodic solution can be established. The analysis turns out to be fairly lengthy, and this suggests that the analysis of more realistic models of electrically active cells (such as the 10-dimensional McAllister–Noble–Tsien model) may present formidable obstacles.

6.2.2.3 *Using entrainment of frequency to describe the primary function of the Purkinje fibers*

In Chapter 4, we described the primary and secondary functions of the Purkinje fiber. The primary function is conduction of the electrical impulses that originate in the pacemaker region and the secondary function is to act as a "backup" pacemaker by "firing" spontaneously and regularly. The periodic solution of system (6.32) that was obtained in the previous section can be regarded as a mathematical description of the secondary function.

Now we turn to a mathematical description of the primary function, that is, a description of the conduction of the electrical impulses.

In view of the lengthy effort that was required to obtain the description of the secondary function, we might expect that the mathematical description of the primary function would require considerable work. Actually the description is derived very easily by applying the theory of entrainment of frequency, and we begin by summarizing this theory briefly.

The physical phenomenon of entrainment of frequency can be described as follows. Suppose a physical system (e.g., a tuning fork) has a natural frequency of oscillation, say ω_0, and suppose an outside force with frequency ω is impressed on the system. If ω is not close to ω_0, that is, if $|\omega - \omega_0|$ is large enough, the two frequencies together produce a beat frequency. This can be observed in the simple experiment of striking two tuning forks, one with frequency ω_0 and one with frequency ω. The oscillations of the nearby air are produced by the sum of the two oscillations, and the experimenter hears the beat frequency. But if the frequency ω is close enough to ω_0, that is, if $|\omega - \omega_0|$ is small enough, then the physical system oscillates with frequency ω. The "free" or natural frequency ω_0 is said to have been "entrained" by the outside or extraneous frequency ω.

Examples of entrainment of frequency have long been observed by scientists. In the seventeenth century, Huyghens observed a mechanical example: Two clocks slightly out of synchronization become synchronized when fixed on a thin wooden board. An example that was once of great practical importance and that motivated the mathematical analysis carried out by van der Pol, was the three-element radio tube (or electron tube) in which the frequency of oscillation of the tube is entrained by the frequency impressed on the grid. Since radio tubes have long since been supplanted by transistors, it is not purposeful to describe this system in detail. [For a mathematical discussion, see Stoker (1950,

p. 147ff) and Minorsky (1962, p. 429ff).] The importance of this example today is that it gave rise to a mathematical theory that has an independent identity and that can be applied to entirely different physical systems such as the Purkinje fiber.

The mathematical description of entrainment of frequency consists, first, of an autonomous system of nonlinear ordinary differential equations that describe the physical system that has an oscillation of frequency ω_0. Let us write this system as

$$\frac{dx}{dt} = f(x), \tag{6.33}$$

where x and f are n vectors. Then the oscillation of frequency ω_0 is a solution $\bar{x}(t)$ of (6.33) such that $\bar{x}(t)$ has period $1/\omega_0$. The postulated outside periodic force is represented by a term that has frequency ω or period $1/\omega$. If we denote this term by a function $g(t)$, then the description of the system and the impressed forcing term is given by

$$\frac{dx}{dt} = f(x) + g(t). \tag{6.34}$$

A mathematical description of entrainment of frequency would be a solution of (6.34) that has period $1/\omega$ and that is "close" (in some sense) to the given solution $\bar{x}(t)$. But the question of whether such a solution exists is a special case of the theory discussed in Chapter 5, where Theorem 11 gives sufficient conditions for the existence. At this stage, we simply emphasize that the entrainment of frequency solution exists independent of the magnitude of $g(t)$. Thus, a forcing term of very small magnitude may produce an oscillation of frequency ω that has a large magnitude.

Now we have already shown that the Purkinje fiber can be regarded as a system with a natural frequency (the frequency of the spontaneous firings) and if we describe the electrical impulses that arrive regularly at any given position on the Purkinje fiber by a periodic forcing term added to the system (6.15), then, in general, the theory of entrainment of frequency will guarantee that the Purkinje fiber will oscillate or fire with a frequency equal to the frequency of the forcing term.

Now we describe this application of entrainment of frequency to the system (6.19):

$$\frac{dV}{dt} = \frac{1}{\varepsilon}\mathscr{F}(V, m, h, n),$$

$$\frac{dm}{dt} = \frac{1}{\varepsilon}\left[\frac{m_\infty(V) - m}{M_m T_m(V)}\right],$$

$$\frac{dh}{dt} = \frac{h_\infty(V) - h}{M_m T_h(V)},$$

$$\frac{dn}{dt} = \eta\left[\frac{n_\infty(V) - n}{M_m T_n(V)}\right].$$

Let us assume that if ε and y are small enough positive numbers, system (6.19) has a nontrivial periodic solution of period, say T_0. Denote this periodic solution by $(\overline{V}(t), \overline{m}(t), \overline{h}(t), \overline{n}(t))$ and write the forcing term that describes the influence of the electrical impulses that arrive regularly from the pacemaker region by $(f_1(t, p), f_2(t, p), f_3(t, p), f_4(t))$, where for $i = 1, 2, 3, 4$, the function $f_i(t, p)$ has continuous first partial derivatives in t and p, and p is a small positive parameter. We assume that $f_i(t, p)$ has period $T(p)$ in t, where $T(p)$ is a continuous real-valued function of p such that $\lim_{p \to 0} T(p)$ exists and

$$\lim_{p \to 0} T(p) = T_0.$$

Finally, we assume that $\lim_{p \to 0} f_i(t, p) = 0$. We consider the question of whether the system

$$\frac{dV}{dt} = \frac{1}{\varepsilon}\mathscr{F}(V, m, h, n) + f_1(t, p),$$

$$\frac{dm}{dt} = \frac{1}{\varepsilon}\left[\frac{m_\infty(V) - m}{M_m T_m(V)}\right] + f_2(t, p),$$

$$\frac{dh}{dt} = \frac{h_\infty(V) - h}{M_m T_h(V)} + f_3(t, p),$$

$$\frac{dn}{dt} = \eta\left[\frac{n_\infty(V) - n}{M_m T_h(V)}\right] + f_4(t, p),$$

has a solution of period $T(p)$ such that, as $p \to 0$, this solution approaches the periodic solution $(\bar{V}(t), \bar{m}(t), \bar{h}(t), \bar{n}(t))$. But this is a special case of the question answered by the theorem of Poincaré (Chap. 5, Theorem 11). Poincaré's theorem states that if a certain hypothesis is satisfied, then a periodic solution exists. The hypothesis concerns the nonexistence of periodic solutions of the linearized equation. This hypothesis can be expressed as the condition that a certain determinant is nonzero, that is, a condition of genericity is satisfied. Since the terms in system (6.32) are, at best, good approximations, we may certainly assume that such a condition is satisfied. Hence, the periodic solution exists and it describes the oscillatory behavior of the Purkinje fiber at any given point along the fiber where the period of the oscillation is the period of the regular behavior of the electrical impulses arriving from the pacemaker region.

Of course our mathematical analysis is far from complete. The most serious gap is the absence of a stability analysis. The periodic solution $(\bar{V}(t), \bar{m}(t), \bar{h}(t))$ should be globally asymptotically stable or at least have a good-sized region of asymptotic stability. Also the periodic solution given by Theorem 11 of Chapter 5 should have a good-sized region of asymptotic stability. Finally, it is important to show that the entrainment of frequency takes place even if $T(p)$ decreases very rapidly, that is, even if dT/dp at $p = 0$ is very negative. The reason for this is that the natural frequency of firing of the Purkinje fiber is often much slower than the pacemaker frequency.

6.2.2.4 *Further directions for study of the Noble model and other cardiac models*

In the preceding sections, we have carried out a detailed analysis of one of the basic mathematical problems that arises in the study of the Noble model. In this section we will describe briefly several other mathematical problems that arise naturally in the study of cardiac models. Since little or no work has been done on most of these problems, we will merely state the problems and, in a few places, indicate possible approaches to the solution of the problems. It should be emphasized that the brevity of these descriptions is a poor measure of the importance or difficulty of the problems.

Computation of period of periodic solution. The natural frequency of the spontaneous oscillations of the Purkinje fiber is known from laboratory observations. Consequently, if a periodic solution of a model such as the Noble model is determined, the next step is to determine the period of the solution and compare this result with the experimental result. Suppose the existence of a periodic solution is established as in the earlier discussion, that is, by finding a closed discontinuous solution and then proving that there is a periodic solution near the closed discontinuous solution. Then one possible way to estimate the periods of the solution is to estimate the time interval required to "travel around" the discontinuous solution. In making this estimate, we assume that travel along the segments of the discontinuous interval that are described by the fast system is instantaneous. Thus to estimate the time interval required to travel around the discontinuous solution is to estimate the time required to travel along the segments of the discontinuous solution that are governed by the slow system. Since the slow system is explicitly known, such an estimate can be carried out by straightforward methods.

Description of the influence of drugs. As pointed out earlier, one of the main purposes in studying mathematical models of electrically active cells is to obtain a clearer and more accurate picture of how the cell functions. In the earlier discussion, we were referring to the normal functioning of the nerve cell or Purkinje fiber. But the study of abnormal functioning, owing, for example, to the influence of drugs, is also very important, especially in cardiac models.

The first step in such a study of abnormal functioning is to derive a mathematical model that takes into account the presence of agents such as adrenaline, tetrodotoxin, or manganese ions. There are many experimental studies in this direction. For work up to 1979, see Noble (1979), especially Chapter 8.

Description of other oscillations in cardiac models. In the earlier discussion, we considered only the normal spontaneous oscillations of the Purkinje fiber. There is another oscillation of the Purkinje fiber that occurs at low membrane potential. This oscillation has been studied experimentally by Hauswirth, Noble, and Tsien (1969). In a sufficiently realistic mathematical model of the Purkinje fiber, a stable periodic solution that describes this oscillation should occur.

Beeler and Reuter (1977) describe experimental conditions under which the myocardial fiber displays oscillatory behavior. Beeler and Reuter show by numerical analysis that their mathematical model of the myocardial fiber (described in detail in Chapter 4 of this book) has oscillatory solutions if terms describing these experimental conditions are added to the original mathematical model. Qualitative analyses of these modifications of the model might show how these oscillations arise and are maintained.

6.2.2.5 *Description of traveling waves in nerve conduction*

In Chapter 2, Section 2.5, we derived the full Hodgkin–Huxley equations, which can be used to describe the normal physiological functioning of the nerve axon. The full Hodgkin–Huxley equations are the partial differential equations

$$\frac{1}{R}\frac{\partial^2 V}{\partial x^2} = C\frac{\partial V}{\partial t} + \bar{g}_{Na}m^3h(V - V_{Na})2\pi rL$$

$$+ \left\{ \bar{g}_K n^4(V - V_K) + \bar{g}_l(V - V_l) \right\}2\pi rL,$$

$$\frac{\partial m}{\partial t} = \alpha_m(1 - m) - \beta_m m, \qquad\qquad (\mathscr{H}\text{-}\mathscr{H})$$

$$\frac{\partial h}{\partial t} = \alpha_h(1 - h) - \beta_h h,$$

$$\frac{\partial n}{\partial t} = \alpha_n(1 - n) - \beta_n n.$$

As pointed out earlier, the general problem of solving the system $(\mathscr{H}\text{-}\mathscr{H})$ is extremely difficult (see the discussion at the end of Chapter 5). However, we are interested only in solutions of $(\mathscr{H}\text{-}\mathscr{H})$ that would describe an action potential. As shown in Chapter 2, these are traveling wave solutions, and to find such solutions, we need to solve the system of ordinary differential equations:

$$\frac{1}{R}\frac{d^2V}{d\xi^2} = C(-\theta)\frac{dV}{d\xi} + \left\{\bar{g}_{Na}m^3h(V - V_{Na})\right\}2\pi rL$$

$$+ \left\{\bar{g}_K n^4(V - V_K) + g_l(V - V_l)\right\}2\pi rL,$$

$$\frac{dm}{d\xi} = -\frac{1}{\theta}\left[\alpha_m(1 - m) - \beta_m m\right], \tag{2.44}$$

$$\frac{dh}{d\xi} = -\frac{1}{\theta}\left[\alpha_h(1 - h) - \beta_h h\right],$$

$$\frac{dn}{d\xi} = -\frac{1}{\theta}\left[\alpha_n(1 - n) - \beta_n n\right].$$

In these equations, θ is a parameter that represents the velocity of the action potential.

As discussed in Chapter 2, the mathematical problem of finding a traveling wave solution becomes the question of how to determine a value of θ for which there are solutions ($V(\xi)$, $m(\xi)$, $h(\xi)$, $n(\xi)$) of (2.44) such that $\lim_{\xi \to \infty} V(\xi) = 0$. Since system (2.44) is a nonlinear system, this question is an example of a nonlinear eigenvalue problem. Numerous studies of this nonlinear eigenvalue problem have been carried out and we will restrict ourselves here to a very brief listing of some of this work.

Before listing the work, however, it seems purposeful to indicate a little of the physiological significance of this nonlinear eigenvalue problem. First of all, as discussed in Chapter 2, Hodgkin and Huxley made a numerical study of the problem in their original work. They found numerically a solution of the problem, that is, they found numerically a value of θ and a corresponding solution ($V(\xi)$, $m(\xi)$, $h(\xi)$, $n(\xi)$) of (2.44) such that $\lim_{\xi \to \infty} V(\xi) = 0$. The value of θ that Hodgkin and Huxley determined represented an estimate of the velocity of the nerve impulse. This velocity had long since been determined in the laboratory. As we pointed out earlier, the comparatively close agreement of θ ($= 18.8$ m/s) and the experimentally observed velocity of 21.2 m/s is one of the most spectacular triumphs of the Hodgkin–Huxley theory. Thus, we have an example of the crucial physiological importance of the nonlinear eigenvalue problem.

On the other hand, in deriving the full Hodgkin–Huxley equations, we take a step away from the reality of experimental data. As it turns out, this step is not significant in the Hodgkin–Huxley work. This is shown, of course, by the successful estimation of the velocity of the nerve impulse. But that same step away from the reality of the laboratory may cause serious difficulties in the study of other electrically active cells. For example, McAllister, Noble, and Tsien (1975) point out that there seem to be serious difficulties in using their model of the cardiac Purkinje fiber to study conduction of impulses along the fiber.

The decision to search for traveling wave solutions may also be an obstacle to realistic study of the action potential. The assumption that the action potential can be described by a traveling wave solution seems eminently reasonable and it is well justified in the case of the nerve axon (again by the successful estimate of the velocity of the impulse). Whether this assumption holds for other electrically active cells remains an open question.

As our discussion indicates, the search for traveling wave solutions of the full Hodgkin–Huxley equations has considerable physiological importance. Now we turn to a brief listing of some of the theoretical and qualitative results in this direction.

Summaries of work on traveling wave solutions of the full Hodgkin–Huxley equations may be found in the papers by Scott (1975) and Cohen (1976). Scott also describes results on traveling wave solutions of other models, for example, the full FitzHugh–Nagumo equations. Especially interesting is the study by Rinzel and Keller (1973) of a simplified FitzHugh–Nagumo equation. This is also discussed by Rinzel (1976), and further results, both qualitative and numerical, have been obtained by Rinzel and Terman (1982). Hastings (1974, 1976, 1975/76) has obtained results for classes of equations like the full FitzHugh–Nagumo equations and like the full Hodgkin–Huxley equations. A particularly interesting technique has been used by Carpenter (1977a, b) to study traveling wave solutions of the FitzHugh–Nagumo equations and the

Hodgkin–Huxley equations. Carpenter regards the subject as a problem in singularly perturbed equations and uses the method of isolating blocks.

So far, we have enumerated only studies concerning the existence of traveling wave. Naturally, the question of stability properties of the traveling wave solution arises. Suppose $(V(\xi), m(\xi), h(\xi), n(\xi))$ is a suitable solution of (2.44). Then the actual traveling wave solution of $(\mathscr{H}\text{-}\mathscr{H})$ is $(v(x - \theta t), m(x - \theta t), h(x - \theta t), n(x - \theta t))$. To study the stability of this solution, we must compare, in some sense, to nearby solutions of $(\mathscr{H}\text{-}\mathscr{H})$. That is, we study the stability of the solution of a partial differential equation. The stability theory in Chapter 5 is *not* applicable. In the case just described, we do *not* study the stability of the solution $(v(\xi), m(\xi), h(\xi), n(\xi))$ of system (2.44). For an introduction to this subject, see Scott (1975). For a detailed discussion, see Evans (1972a–c, 1975).

Appendix

In order to understand the experiments of Hodgkin and Huxley, a little background in electricity is needed. We will assume that the reader is familiar with the notions of electrical charge, potential difference, current, resistance, and the units of measure of these. We will also need a basic relation among these, that is, Ohm's law, which is the statement

$$V = IR,$$

where V is the potential difference, I the current, and R the resistance.

In addition, we need to use the notion of capacitance. Since capacitance is a less familiar electrical concept and because it plays a crucial role in electrophysiology, we describe it in some detail.

Let A, B be conducting plates made, for example, of copper. Let A be connected to G, a gold leaf electroscope. If a charge is put on A by touching A with a charged rod, then, because the charge distributes itself uniformly in the gold leaf, the two parts of the gold leaf have the same charge. Thus, they are repelled by one another and hence the leaf is deflected as indicated in Fig. A.1. The amount of deflection measures the potential of A.

Now if plate B, which has no charge, is brought near A as indicated in Fig. A.1, there is a decrease in the deflection of the leaf in G. This is because a rearrangement of electrons and ions in B occurs and this rearrangement neutralizes part of the electric charge on A. For example, if there is a positive charge uniformly distributed on A and G, then electrons will move on B so that they are as close as possible to A. Thus, some of the positive ions on A and G will be bound to A. The result is that the potential on A is decreased even though the electric charge on A remains constant.

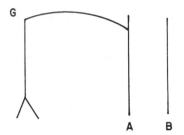

Figure A.1.

Definition. The quantity of electricity (i.e., electrical charge) required to impart unit potential to a conductor is the *capacitance* of the conductor. If q units of electricity raise the potential of the conductor by V units, the capacitance is given by

$$c = \frac{q}{V}.$$

In the experiment just described, the capacitance of A is increased by bringing B closer to A. If an insulating material is placed between A and B, the deflection of the gold leaf is further reduced, that is, the capacitance of A is further increased.

Definition. A pair of conductors carrying equal and opposite quantities of electricity and separated by insulating material is a *capacitor* or *condenser*. The conductors are called the *plates* and the insulating material is called the *dielectric*. The *capacitance C* of a capacitor is given by

$$C = \frac{Q}{V},$$

where Q is the quantity of electricity on either plate and V is the potential difference between the plates.

Capacitance is measured in farads [F = 1 C/V (coulomb/Volt)]. In other words, a capacitor has a capacitance of 1 F if the potential difference between the conductors is changed by 1 V if 1 C of electric charge is transferred from one of the conductors of the capacitor to the other.

The capacitance of a capacitor depends upon: (i) size of the plates (capacitance increases with size of the plates); (ii) distance apart of the plates (capacitance increases when plates are brought closer together); (iii) intervening dielectric. [The influence of the dielectric can be visualized in terms of dipole molecules. When the molecules are oriented properly, they neutralize the charges on the plates, thus decreasing the potential and increasing the capacitance. The measure of the influence of the dielectric is the *permittivity* or *dielectric constant*. If a given capacitor with free space (or air) between the plates has capacitance C and the capacitance increases to C' when some other dielectric is substituted, the permittivity of that dielectric is C'/C.]

Now we consider a simple electrical circuit (Fig. A.2) in which there is a capacitor C, a battery B, and a switch S. If the switch is closed, there is a surge of electrons toward one plate of the capacitor and a surge of electrons away from the other plate and toward the battery terminal. This is not a current in the sense of a continuous flow of electrons or ions through a conductor, but it is a current in the sense that there is movement of electrons. It is sometimes called a *displacement current*.

If a displacement current i_c occurs during a time interval Δt (during which the switch is closed), an electric charge $i_c(\Delta t)$ is added to one plate and removed from the other. A potential difference between the plates, denoted by ΔV, is developed and ΔV is proportional to the charge $Q = i_c(\Delta t)$; that is,

$$Q = i_c(\Delta t) = C(\Delta V),$$

where C is the capacitance of the capacitor. Then

$$C\frac{\Delta V}{\Delta t} = i_c.$$

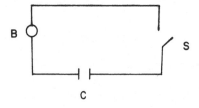

Figure A.2.

Letting $\Delta t \to 0$, we obtain

$$C\frac{\partial V}{\partial t} = i_c.$$

Displacement currents are crucially important in our study because it turns out that the membrane that surrounds the axon behaves as a capacitor. Hence, the total current I across the membrane is

$$I = C_M\frac{\partial V}{\partial t} + I_i, \qquad (A.1)$$

where I_i is the current caused by the flow of ions across the membrane, $C_M(\partial V/\partial t)$ is the displacement current, and C_M is the capacitance per unit area of the membrane.

If we speak more precisely, we should say that it is *assumed* that equation (A.1) is a correct description of the current I. This assumption is one of the basic assumptions in the work of Hodgkin and Huxley.

REFERENCES

Adelman, W. J. and Fitzhugh, R. (1975). Solutions of the H–H equation modified for potassium accumulation in a periaxonal space. *Fed. Amer. Soc. Exp. Biol. Proc.*, 34:1322–1329.

Adelman, W. J. and Palti, Y. (1969). The influence of external potassium on the inactivation of sodium currents in the giant axon of the squid, *Loligo pealei*. *J. Gen. Physiol.*, 53:685–703.

Adelman, W. J. ed. (1971). *Biophysics and Physiology of Excitable Membranes*. Van Nostrand Reinhold, New York.

Adelman, W. J., Palti, Y., and Senft, J. P. (1973). Potassium ion accumulation in a periaxonal space and its effect on the measurement of membrane potassium ion conductance. *J. Membrane Biol.*, 13:387–410.

Adrian, R. H. (1975). Conduction velocity and gating current in the squid giant axon. *Proc. Roy. Soc. Lond. B*, 189:81–86.

Adrian, R. H., Chandler, W. K., and Hodgkin, A. L. (1970). Voltage clamp experiments in striated muscle fibres. *J. Physiol.*, 208:607–644.

Adrian, R. H. and Peachey, L. D. (1973). Reconstruction of the action potential of frog sartorius muscle. *J. Physiol.*, 235:103–131.

Beall, P. T. (1981). The water of life. *The Sciences*, 21:6–9.

Beeler, G. W. and Reuter, H. (1977). Reconstruction of the action potential of ventricular myocardial fibres. *J. Physiol.*, 268:177–210.

Berkinblit, M. B., Dudzyavichus, I., Kovalev, S., Fomin, S. V., Kholopov, A. V., and Chailakhyan, L. M. (1970). Modelling with a computer of the behavior of the membrane of the nerve fibre on rhythmic stimulation. *Biofizikia*, 15:147–155.

Blair, E. A. and Erlanger, J. (1936). On the process of excitation by brief shocks in axons. *Amer. J. Physiol.*, 114:309–316.

Carpenter, G. (1977a). A geometric approach to singular perturbation problems with applications to nerve impulse equations. *J. Differential Equations*, 23:335–367.

Carpenter, G. (1977b). Periodic solutions of nerve impulse equations. *J. Math. Anal. Appl.*, 58:152–173.

Cesari, L. (1971). *Asymptotic Behavior and Stability Problems in Ordinary Differential Equations*. 3rd ed. Berlin: Springer.

Cohen, H. (1976). Mathematical developments in Hodgkin–Huxley theory and its approximations. *Some Mathematical Questions in Biology. VII. Lectures on Mathematics in the Life Sciences*. Vol. 8. Providence, R.I.: American Mathematical Society.

Cole, K. S. (1949). Dynamic electrical characteristics of the squid axon membrane. *Arch. Sci. Physiol.*, 3:253.

Cole, K. S., Antosiewicz, H. A., and Rabinowitz, P. (1955). Automatic computation of nerve excitation. *J. Soc. Ind. Appl. Math.*, 3:153–172.

Cole, K. S. and Moore, J. W. (1961). Potassium ion current in the squid giant axon: dynamic characteristic. *Biophys. J.*, 1:1–14.

Cooley, J., Dodge, F., and Cohen, H. (1965). Digital computer solutions for excitable membrane models. *J. Cell. Comp. Physiol.*, 66:99–108.

Cronin, J. (1964). *Fixed Points and Topological Degree in Nonlinear Analysis.* Mathematical Surveys No. 11. Providence, R.I.: American Mathematical Society.

Cronin, J. (1977). Mathematical aspects of periodic catatonic schizophrenia. *Bull. Math. Biol.*, 39:187–199.

Cronin, J. (1980a). A criterion for asymptotic stability. *J. Math. Anal. Appl.*, 74:247–269.

Cronin, J. (1980b). *Differential Equations, Introduction and Qualitative Theory.* New York: Marcel Dekker.

Cronin, J. (1981). *Mathematics of Cell Electrophysiology. Lecture Notes in Pure and Applied Mathematics.* Vol. 63. New York: Marcel Dekker.

DiFrancesco, D. and Noble, D. (1985). A model of cardiac electrical activity incorporating ionic pumps and concentration changes. *Philos. Trans. Roy. Soc. Lond. B*, 307:353–398.

Dodge, F.A. (1972). On the transduction of visual, mechanical and chemical stimuli. *Internat. J. Neurosci.*, 3:5–14.

Dodge, F. A. and Frankenhaeuser, B. (1959). Sodium currents in the myelinated nerve fibre of *Xenopus laevis* investigated with the voltage clamp technique. *J. Physiol.*, 148:188–200.

Evans, J. W. (1972a). Nerve axon equations. I. Linear approximations. *Indiana Univ. Math. J.*, 21:877–884.

Evans, J. W. (1972b). Nerve axon equations. II. Stability at rest. *Indiana Univ. Math. J.*, 22:7–90.

Evans, J. W. (1972c). Nerve axon equations. III. Stability of the nerve impulse. *Indiana Univ. Math. J.*, 22:577–593.

Evans, J. W. (1975). Nerve axon equations. IV. The stable and the unstable pulse. *Indiana Univ. Math. J.*, 24:1169–1190.

Fife, P. C. (1978). Asymptotic states for equations of reaction and diffusion. *Bull. Amer. Math. Soc.*, 84:693–726.

FitzHugh, R. (1961). Impulses and physiological states in models of nerve membrane. *Biophys. J.*, 1:445–466.

FitzHugh, R. (1965). A kinetic model of the conductance changes in nerve membrane. *J. Cellular Comparative Physiol.*, 66:111–117.

FitzHugh, R. (1969). Mathematical models of excitation and propagation in nerve. In *Biological Engineering.* H. P. Schwan, ed., Chap. 1. New York: McGraw-Hill.

FitzHugh, R. and Antosiewicz, H. A. (1959). Automatic computation of nerve excitation – detailed corrections and additions. *J. Soc. Ind. Appl. Math.*, 7:447–458.

Frankenhaeuser, B. (1962). Instantaneous potassium currents in myelinated nerve fibres of *Xenopus laevis. J. Physiol.*, 160:46–53.

Frankenhaeuser, B. and Hodgkin, A. L. (1956). The after-effects of impulses in the giant nerve fibres of *Loligo. J. Physiol.*, 131:341–376.

Frankenhaeuser, B. and Hodgkin, A. L. (1957). The action of calcium on the electrical properties of squid axons. *J. Physiol.*, 137:218–244.

Frankenhaeuser, B. (1960). Quantitative description of sodium currents in myelinated nerve fibres of *Xenopus laevis*. *J. Physiol.*, 151:491–501.

Frankenhaeuser, B. (1963a). A quantitative description of potassium currents in myelinated nerve fibres of *Xenopus laevis*. *J. Physiol.*, 169:424–430.

Frankenhaeuser, B. (1963b). Inactivation of the sodium-carrying mechanism in myelinated nerve fibres of *Xenopus laevis*. *J. Physiol.*, 169:445–451.

Frankenhaeuser, B. and Huxley, A. F. (1964). The action potential in the myelinated nerve fibre of *Xenopus laevis* as computed on the basis of voltage clamp data. *J. Physiol.*, 171:302–315.

Fricke, H. (1923). The electric capacity of cell suspensions. *Phys. Rev.*, 21:708–709.

Goldman, D. E. (1943/44). Potential, impedance, and rectifications in membranes. *J. Gen. Physiol.*, 27:37–60.

Goldman, L. and Schauf, C. L. (1972). Inactivation of the sodium current in *Myxicola* giant axons. Evidence for coupling to the activation process. *J. Gen. Physiol.*, 59:659–675.

Guttman, R., Lewis, S., and Rinzel, J. (1980). Control of repetitive firing in squid axon membrane as a model for a neuronoscillator. *J. Physiol.*, 305:377–395.

Hagiwara, S. and Oomura, Y. (1958). The critical depolarization for the spike in the squid giant axon. *Japan J. Physiol.*, 8:234–245.

Hahn, W. (1967). *Stability of Motion*. New York: Springer.

Hale, J. K. (1963). *Oscillations in Nonlinear Systems*. New York: McGraw-Hill.

Hastings, S. (1974). The existence of periodic solutions to Nagumo's equation. *Quart. J. Math.*, 25:369–378.

Hastings, S. (1975/76). On traveling wave solutions of the Hodgkin–Huxley equations. *Arch. Rational Mech. Anal.*, 60:229–257.

Hastings, S. (1976). On the existence of homoclinic and periodic orbits for the FitzHugh–Nagumo equations. *Quart. J. Math.*, 27:123–134.

Hauswirth, O., Noble, D., and Tsien, R. W. (1969). The mechanism of oscillatory action at low membrane potentials in cardiac Purkinje fibers. *J. Physiol.*, 200:255–265.

Hodgkin, A. L. (1951). The ionic basis of electrical activity in nerve and muscle. *Biol. Rev.*, 26:339–409.

Hodgkin, A. L. (1971). *The Conduction of the Nervous Impulse*. Liverpool: Liverpool University Press.

Hodgkin, A. L. (1975). The optimum density of sodium channels in an unmyelinated nerve. *Philos. Trans. Roy. Soc. Lond. B*, 270:297–300.

Hodgkin, A. L. (1977). Chance and design in electrophysiology: an informal account of certain experiments on nerve carried out between 1934 and 1952. *The Pursuit of Nature: Informal Essays on the History of Physiology*. Cambridge: Cambridge University Press.

Hodgkin, A. L. and Huxley, A. F. (1952a). Currents carried by sodium and potassium ions through the membrane of the giant axon of *Loligo*. *J. Physiol.*, 116:449–472.

Hodgkin, A. L. and Huxley, A. F. (1952b). The components of membrane conductance in the giant axon of *Loligo*. *J. Physiol.*, 116:473–496.

Hodgkin, A. L. and Huxley, A. F. (1952c). The dual effect of membrane potential on sodium conductance in the giant axon of *Loligo*. *J. Physiol.*, 116:497–506.

Hodgkin, A. L. and Huxley, A. F. (1952d). A quantitative description of membrane and its application to conduction and excitation in nerve. *J. Physiol.*, 117:500–544.

Hodgkin, A. L., Huxley, A. F., and Katz, B. (1952). Measurement of current-voltage relations in the membrane of the giant axon of *Loligo*. *J. Physiol.*, 116:424–448.

Hodgkin, A. L. and Katz, B. (1949). The effect of sodium ions on the electrical activity of the giant axon of the squid. *J. Physiol.*, 108:33–77.

Hoyt, R. C. (1963). The squid giant axon, mathematical models. *Biophys. J.*, 3:399–431.

Hoyt, R. C. (1968). Sodium inactivation in nerve fibers. *Biophys. J.*, 8:1074–1097.

Hoyt, R. C. and Adelman, W. J. (1970). Sodium inactivation, experimental test of two models. *Biophys. J.*, 10:610–617.

Hunter, P. J., McNaughton, P. A., and Noble, D. (1975). Analytical models of propagation in excitable cells. *Progr. Biophys. Molecular Biol.*, 30:99–144.

Hurewicz, W. (1958). *Lectures on Ordinary Differential Equations*. New York: Wiley.

Huxley, A. F. (1959). Ion movements during nerve activity. *Ann. N. Y. Acad. Sci.*, 51:221–246.

Jack, J. J. B., Noble, D., and Tsien, R. W. (1975). *Electric Current Flow in Excitable Cells*. Oxford: Clarendon Press.

Jakobsson, E. (1973). The physical interpretation of mathematical models for sodium permeabiity changes in excitable membranes. *Biophys. J.*, 13:1200–1211.

Jakobsson, E. and Guttman, R. (1980). The standard Hodgkin–Huxley model and squid axons in reduced external Ca^{++} fail to accommodate to slowly rising currents. *Biophys. J.*, 31:293–298.

Krasnosel'skii, M. A. (1964). *Topological Methods in the Theory of Nonlinear Integral Equations* (translated from Russian). Macmillan, New York.

LaSalle, J. and Lefshetz, S. (1961). *Stability by Liapunov's Direct Method With Applications*. New York: Academic.

Lecar, H. and Nossal, R. (1971). Theory of threshold fluctuations in nerves. I. Relationships between electrical noise and fluctuations in axon firing. *Biophys. J.*, 11:1048–1067.

Lefschetz, S. (1963). *Differential Equations: Geometric Theory*. 2nd ed. New York: Interscience.

Levinson, N. (1951). Perturbations of discontinuous solutions of non-linear systems of differential equations. *Acta Math.* 82:71–106.

Lloyd, N. G. (1978). *Degree Theory*. Cambridge: Cambridge University Press.

Luria, S. E. (1975). *36 Lectures in Biology*. Cambridge, Mass.: The MIT Press.

Macaulay, F. S. (1916). *The Algebraic Theory of Modular Systems*. Cambridge: Cambridge University Press.

Marden, M. (1966). *Geometry of Polynomials*. Mathematical Surveys No. 3. Providence, R.I.: American Mathematical Society.

McAllister, R. E., Noble, D., and Tsien, R. W. (1975). Reconstruction of the electrical activity of cardiac Purkinje fibers. *J. Physiol.*, 251:1–58.

McDonough, J. (1979). Doctoral thesis, Rutgers University.

McKean, H. P. (1970). Nagumo's equation. *Advan. Math.*, 4:209–223.

Minorsky, N. (1962). *Nonlinear Oscillations*. Princeton, N.J.: Van Nostrand.

Mishchenko, E. F. and Rozov, N. Kh. (1980). *Differential Equations With Small Parameters and Relaxation Oscillations*. Translated from Russian by F. M. C. Goodspeed. New York: Plenum.

Nagumo, J. S., Arimoto, S., and Yoshizawa, S. (1962). An active pulse transmission line simulating nerve axon. *Proc. IRE*, 50:2061–2070.

Neher, E. and Stevens, C. F. (1977). Conductance fluctuations and ionic pores in membranes. *Ann. Rev. Biophys. Bioeng.*, 6:345–381.

Nemytskii, V. V. and Stepanov, V. V. (1960). *Qualitative Theory of Differential Equations*. Princeton, N.J.: Princeton University Press.

Noble, D. (1962). A modification of the Hodgkin–Huxley equations applicable to Purkinje fibre action and pace-maker potentials. *J. Physiol.*, 160: 317–352.

Noble, D. (1974). Cardiac action potentials and pacemaker activity. In *Recent Advances in Physiology*. R. J. Linden, ed., Vol. 9, pp. 1–50. Edinburgh: Churchill Livingstone.

Noble, D. (1979). *The Initiation of the Heartbeat*. 2nd ed. Oxford: Oxford University Press.

Poincaré, H. (1881–1934). *Oeuvres*. Tome 1, pp. 3–84; Tome 9, pp. 161, 167–221 (1916–1934); *J. Math. Pures Appl.* (3), 7:375–422 (1881); 8:251–296 (1882); (4), 1:167–244 (1885); 2:151–217 (1886).

Poincaré, H. (1892–99). *Les Méthodes Nouvelles de la Mécanique Céleste*. 3 volumes. Paris: Gauthiers-Villars.

Poore, A. B. (1976). On the theory and application of the Hopf–Friedrichs bifurcation theory. *Arch. Rational Mech. Anal.*, 60:371–393.

Rinzel, J. (1976). Simple model equations for active nerve conduction and passive neuronal integration. *Some Mathematical Questions in Biology. VII. Lectures on Mathematics in the Life Sciences*. Vol. 8. Providence, R.I.: American Mathematical Society.

Rinzel, J. (1979). On repetitive activity in nerve. *Fed. Proc.*, 37:2793–2802.

Rinzel, J. and Keller, J. B. (1973). Traveling wave solutions of a nerve conduction equation. *Biophys. J.*, 13:1313–1337.

Rinzel, J. and Terman, D. (1982). Propagation phenomena in a bistable reaction diffusion system. *SIAM J. Appl. Math.*, 42:1111–1137.

Schwartz, T. L. (1971). The thermodynamic foundations of membrane physiology. In *Biophysics and Physiology of Excitable Membranes*. W. J. Adelman, ed. New York: Van Nostrand Reinhold.

Scott, A. C. (1975). The electrophysics of a nerve fiber. *Rev. Mod. Phys.*, 47:487–533.

Sell, G. (1966). Periodic solutions and asymptotic stability. *J. Differential Equations*, 2:143–157.

Sibuya, Y. (1960). On perturbations of discontinuous solutions of ordinary differential solutions. *Natur. Sci. Rep. Ochanomizu Univ.*, 11:1–18.

Stoker, J. J. (1950). *Nonlinear Vibrations*. New York: Interscience.

Tasaki, I. (1956). Initiation and abolition of the action potential of a single node of Ranvier. *J. Gen. Physiol.*, 39:377–395.

Ten Hoopen, M. and Verveen, A. A. (1963). Nerve-model experiments on fluctuation in excitability. *Nerve, Brain and Memory Models, Progress in Brain Research*. N. Wiener and J. P. Schade, eds., Vol. 2. Amsterdam: Elsevier.

Tille, J. (1965). A new interpretation of the dynamic changes of the potassium conductance in the squid giant axon. *Biophys. J.*, 5:163–176.

Troy, W. C. (1974/75). Oscillation phenomena in the Hodgkin–Huxley equations. *Proc. Roy. Soc. Edinburgh*, 74A:299–310.

van der Waerden, B. (1940). *Modern Algebra*. Vol. II, 2nd ed. Berlin: Springer.

Verveen, A. A. and Derksen, H. E. (1968). Fluctuating phenomena in nerve membrane. *Proc. IEEE*, 56:906–916.

Worden, F. G., Swazey, J. P., and Adelman, G., eds. (1975). *The Neurosciences: Paths of Discovery*. Cambridge, Mass.: MIT Press.

Young, J. Z. (1936). Structure of nerve fibers and synapses in some invertebrates. *Cold Spring Harbor Symp. Quant. Biol.*, 4:1–6.

Zeeman, E. C. (1972). Differential equations for the heartbeat and nerve impulse. In *Towards a Theoretical Biology 4. Essays*. C. H. Waddington, ed., pp. 8–67. Edinburgh: Edinburgh University Press.

Zeeman, E. C. (1973). Differential equations for the heartbeat and nerve impulse. In *Dynamical Systems*. M. M. Peixoto, ed., pp. 683–741. New York: Academic.

Zeeman, E. C. (1977). *Catastrophe Theory: Selected Papers, 1972–1977*. Reading, Mass.: Addison-Wesley.

INDEX